RISK MANAGEMENT SERIES

Safe Rooms and Shelters

PROTECTING PEOPLE AGAINST TERRORIST ATTACKS

FEMA

RISK MANAGEMENT SERIES

Safe Rooms and Shelters

FOREWORD AND ACKNOWLEDGMENTS

OVERVIEW

This manual is intended to provide guidance for engineers, architects, building officials, and property owners to design shelters and safe rooms in buildings. It presents information about the design and construction of shelters in the work place, home, or community building that will provide protection in response to manmade hazards. Because the security needs and types of construction vary greatly, users may select the methods and measures that best meet their individual situations. The use of experts to apply the methodologies contained in this document is encouraged.

The information contained herein will assist in the planning and design of shelters that may be constructed outside or within dwellings or public buildings. These safe rooms will protect occupants from a variety of hazards, including debris impact, accidental or intentional explosive detonation, and the accidental or intentional release of a toxic substance into the air. Safe rooms may also be designed to protect individuals from assaults and attempted kidnapping, which requires design features to resist forced entry and ballistic impact. This covers a range of protective options, from low-cost expedient protection (what is commonly referred to as sheltering-in-place) to safe rooms ventilated and pressurized with air purified by ultra-high-efficiency filters. These safe rooms protect against toxic gases, vapors, and aerosols (finely divided solid or liquid particles). The contents of this manual supplement the information provided in FEMA 361, *Design and Construction Guidance for Community Shelters* and FEMA 320, *Taking Shelter From the Storm: Building a Safe Room Inside Your House.* In conjunction with FEMA 361 and FEMA 320, this publication can be used for the protection of shelters against natural disasters. Although this publication specifically does not address nuclear explosions and shelters that protect against radiological fallout, that information may be found in FEMA TR-87, *Standards for Fallout Shelters.*

This guidance focuses on safe rooms as standby systems, ones that do not provide protection on a continuous basis. To employ a standby system requires warning based on knowledge that a hazardous condition exists or is imminent. Protection is initiated as a result of warnings from civil authorities about a release of hazardous materials, visible or audible indications of a release (e.g., explosion or fire), the odor of a chemical agent, or observed symptoms of exposure in people. Although there are automatic detectors for chemical agents, such detectors are expensive and limited in the number of agents that can be reliably detected. Furthermore, at this point in time, these detectors take too long to identify the agent to be useful in making decisions in response to an attack. Similarly, an explosive vehicle or suicide bomber attack rarely provides advance warning; therefore, the shelter is most likely to be used after the fact to protect occupants until it is safe to evacuate the building.

Two different types of shelters may be considered for emergency use, standalone shelters and internal shelters. A standalone shelter is a separate building (i.e., not within or attached to any other building) that is designed and constructed to withstand the range of natural and manmade hazards. An internal shelter is a specially designed and constructed room or area within or attached to a larger building that is structurally independent of the larger building and is able to withstand the range of natural and manmade hazards. Both standalone and internal shelters are intended to provide emergency refuge for occupants of commercial office buildings, school buildings, hospitals, apartment buildings, and private homes from the hazards resulting from a wide variety of extreme events.

The shelters may be used during natural disasters following the warning that an explosive device may be activated, the discovery of an explosive device, or until safe evacuation is established following the detonation of an explosive device or the release of a toxic substance via an intentional aerosol attack or an industrial accident. Standalone community shelters may be constructed in neighborhoods where existing homes lack shelters. Community shelters may be intended for use by the occupants of buildings they are constructed within or near, or they may be intended for use by the residents of surrounding or nearby neighborhoods or designated areas.

BACKGROUND

The attack against the Alfred P. Murrah Federal Office Building in Oklahoma City and the anthrax attacks in October 2001 made it clear that chemical, biological, radiological, and explosive (CBRE) attacks are a credible threat

> For additional information on CBR and explosives, see FEMA 426 and other Risk Management Series publications.

to our society. Such attacks can cause a large number of fatalities or injuries in high-occupancy buildings (e.g., school buildings, hospitals and other critical care facilities, nursing homes, day-care centers, sports venues, theaters, and commercial buildings) and residential neighborhoods.

Protection against the effects of accidental or intentional explosive detonations and accidental or intentional releases of toxic substances into the air or water represent a class of manmade hazards that need to be addressed along with the protection that may already be provided against the effects of natural hazards such as hurricanes and tornadoes. Although there are a wide range of scenarios that may create these manmade hazards, to date they are extremely rare events. However, although scarce, these events warrant consideration for passive protective measures. These passive protective measures may be in the form of a safe room in which occupants of a building may be sheltered until it is safe to evacuate. The effectiveness of the safe room for protecting occupants from manmade threats is dependent on the amount of warning prior to the event and its construction. For example, in Israel, a building occupant may expect a 3-minute warning prior to a Scud missile attack; therefore, the shelter must be accessible to all building occupants within this time period. Note that such advance warning rarely accompanies the explosive vehicle or suicide bomber event; in this case, the function of the safe room is to protect occupants until law enforcement agencies determine it is safe to evacuate.

Protection against explosive threats depends to a great extent on the size of the explosive, the distance of the detonation relative to the shelter, and the type of construction housing the shelter. Although there may be opportunities to design a new facility to protect against a specified attack scenario, this may be of limited feasibility for the retrofit of an existing building. The appropriate combination of charge weight and standoff distance as well as the intervening structure between the origin of threat and the protected space is very site-specific; therefore, it is impractical to define a design level threat in these terms. Rather than identify a shelter to resist a specified explosive threat, this document will provide guidance that will address different types of building construction and the reasonable measures that may be taken to provide a secure shelter and a debris mitigating enclosure for a shelter. This approach does not attempt to address a specific threat because there are too many possible scenarios to generalize a threat-specific approach; however, it does allow the user to determine the feasible options that may be evaluated on a case by case basis to determine a response to any postulated threat. For protection against assault and attempted kidnapping, a level of forced entry and ballistic resistance may be specified. Several different organizations (e.g., the American Society for Testing and Materials (ASTM), H.P. White, Underwriters Laboratories (UL), the Department of Justice (DOJ), etc.) define performance levels associated with forced entry and ballistic resistance that relate to the different sequence of tests that are required to demonstrate effectiveness of a given construction product. This document will not distinguish between the different types of testing regimes.

Protection against airborne hazardous materials may require active measures. Buildings are designed to exchange air with the outdoors in normal operation; therefore, airborne hazardous materials can infiltrate buildings readily when released outdoors, driven by pressures generated by wind, buoyancy, and fans. Buildings also tend to retain contaminants; that is, it takes longer for the toxic materials to be purged from a building than to enter it.

The safe room may also shelter occupants from tornadoes and hurricanes, which are the most destructive forces of nature. Since 1995, over 1,200 tornadoes have been reported nationwide each year. Approximately five hurricanes strike the United States mainland every 3 years and two of these storms will cause extensive damage. Protection from the effects of these natural occurrences may be provided by well designed and amply supplied safe rooms. The well designed safe room protects occupants from the extremely rare, but potentially catastrophic effects of a manmade threat as well as the statistically more common, but potentially less severe effects of a natural disaster.

SCOPE AND ORGANIZATION OF THE MANUAL

This document will discuss the design of shelters to protect against CBRE attacks. Fallout shelters that are designed to protect against the effects of a nuclear weapon attack are not addressed in this publication. The risks of death or injury from CBRE attacks are not evenly distributed throughout the United States. This manual will guide the reader through the process of designing a shelter to protect against CBRE attacks. The intent of this manual is not to mandate the construction of shelters for CBRE events, but rather to provide design guidance for persons who wish to design and build such shelters.

The design and planning necessary for extremely high-capacity shelters that may be required for large, public use venues such as stadiums or amphitheaters are beyond the scope of this design manual. An owner or operator of such a venue may be guided by concepts presented in this document, but detailed guidance concerning extremely high-capacity shelters is not provided. The design of such shelters requires attention to issues such as egress and life safety for a number of people that are orders of magnitude greater than those proposed for a shelter designed in accordance with the guidance provided herein.

The intent of this manual is not to override or replace current codes and standards, but rather to provide important guidance of best practices (based on current technologies and scientific research) where none has been available. No known building, fire, life safety code, or engineering standard has previously attempted to provide detailed information, guidance, and recommendations concerning the design of CBRE shelters for protection of the general public. Therefore, the information provided herein is the best available at the time this manual was published. Designing and constructing a shelter according to the criteria in this manual does not mean that the shelter will be capable of withstanding every possible event. The design professional who ultimately designs a shelter should state the limiting assumptions and shelter design parameters on the project documents.

This manual includes the following chapters and appendices:

○ Chapter 1 presents design considerations, potential threats, the levels of protection, shelter types, siting, occupancy duration, and human factors criteria for shelters (e.g., square footage per shelter occupant, proper ventilation, distance/travel time and accessibility, special needs, lighting, emergency power, route marking and wayfinding, signage, evacuation considerations, and key operations zones).

○ Chapter 2 discusses the structural design criteria for blast and impact resistance, as well as shelters and model building types. Structural systems and building envelope elements for shelters are analyzed and protective design measures for the defined building types are provided.

○ Chapter 3 describes how to add chemical, biological, and radiological (CBR) protection capability to a shelter or a safe room. It also discusses air filtration, safe room criteria, design requirements, operations and maintenance, commissioning, and training required to operate a shelter.

○ Chapter 4 discusses emergency management considerations, Federal CBRE response teams, emergency response and mass care, community shelter operations plans, descriptions of the responsibilities of the shelter team members, shelter equipment and supplies, maintenance plans, and commercial building shelter operation plans. Key equipment considerations and training are also discussed.

○ Appendix A presents the references used in the preparation of this document.

○ Appendix B contains a list of acronyms and abbreviations that appear in this document.

ACKNOWLEDGMENTS

Principal Authors:

Robert Smilowitz, Weidlinger Associates Inc.

William Blewett, Battelle Memorial Institute

Pax Williams, Battelle Memorial Institute

Michael Chipley, PBS&J

Contributors:

Milagros Kennett, FEMA, Project Officer, Risk Management Series Publications

Eric Letvin, URS, Project Manager

Deb Daly, Greenhorne & O'Mara, Inc.

Julie Liptak, Greenhorne & O'Mara, Inc.

Wanda Rizer, Consultant

Project Advisory Panel:

Ronald Barker, DHS, Office of Infrastructure Protection

Wade Belcher, General Service Administration

Curt Betts, U.S. Army Corps of Engineers

Robert Chapman, NIST

Ken Christenson, U.S. Army Corps of Engineers

Roger Cundiff, DOS

Michael Gressel, CDC, NIOSH

Marcelle Habibion, Department of Veterans Affairs

Richard Heiden, U.S. Army Corps of Engineers

Nancy McNabb, NFPA

Kenneth Mead, CDC, NIOSH

Arturo Mendez, NYPD/DHS Liaison

Rudy Perkey, NAVFAC

Joseph Ruocco, SOM

Robert Solomon, NFPA

John Sullivan, PCA

TABLE OF CONTENTS

CHAPTER 2 – STRUCTURAL DESIGN CRITERIA

CHAPTER 3 – CBR THREAT PROTECTION

APPENDICES

TABLES

FIGURES

Chapter 2

Chapter 3

Chapter 4

1.1 OVERVIEW

The attack against the Alfred P. Murrah Federal Office Building in Oklahoma City and the anthrax attacks in October 2001 made it clear that chemical, biological, radiological, and explosive (CBRE) attacks are a credible threat to our society. Such attacks can cause a large number of fatalities or injuries in high-occupancy buildings (e.g., school buildings, hospitals and other critical care facilities, nursing homes, day-care centers, sports venues, theaters, and commercial buildings) and residential neighborhoods.

This chapter discusses the potential manmade threats to which a shelter may be exposed and the level of protection (LOP) that may be assumed by building owners when deciding to build a shelter to support the preparedness objectives established in the National Preparedness Goal. This guidance complements other shelter publications such as the American Red Cross (ARC) 4496, *Standards for Hurricane Evacuation Shelter Selection*; FEMA 320, *Taking Shelter From the Storm: Building a Safe Room Inside Your House*; and FEMA 361, *Design and Construction Guidance for Community Shelters.*

This manual presents information about the design and construction of shelters in the work place, home, or community building that will provide protection

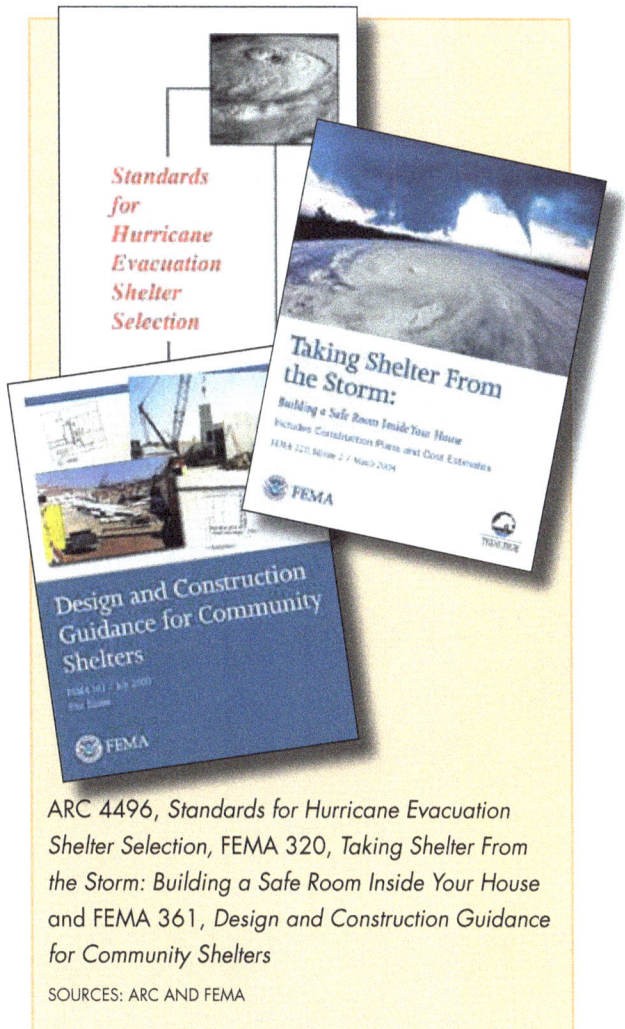

ARC 4496, *Standards for Hurricane Evacuation Shelter Selection*, FEMA 320, *Taking Shelter From the Storm: Building a Safe Room Inside Your House* and FEMA 361, *Design and Construction Guidance for Community Shelters*

SOURCES: ARC AND FEMA

in response to the manmade CBRE threats as defined in the National Response Plan (NRP) and the National Planning Scenarios. As published in the *National Preparedness Guidance* (April 2005), the Federal interagency community developed 15 planning scenarios (the National Planning Scenarios or Scenarios) for use in national, Federal, state, and local homeland security preparedness activities. The National Planning Scenarios are planning tools and are representative of the range of potential terrorist attacks and natural disasters and the related impacts that face our nation. The scenarios establish the range of response requirements to facilitate preparedness planning.

The National Planning Scenarios describe the potential scope and magnitude of plausible major events that require coordination among various jurisdictions and levels of government and communities.

Scenario 1: Nuclear Detonation – 10-Kiloton Improvised Nuclear Device

Scenario 2: Biological Attack – Aerosol Anthrax

Scenario 3: Biological Disease Outbreak – Pandemic Influenza

Scenario 4: Biological Attack – Plague

Scenario 5: Chemical Attack – Blister Agent

Scenario 6: Chemical Attack – Toxic Industrial Chemicals

Scenario 7: Chemical Attack – Nerve Agent

Scenario 8: Chemical Attack – Chlorine Tank Explosion

Scenario 9: Natural Disaster – Major Earthquake

Scenario 10: Natural Disaster – Major Hurricane

Scenario 11: Radiological Attack – Radiological Dispersal Devices

Scenario 12: Explosives Attack – Bombing Using Improvised Explosive Device

Scenario 13: Biological Attack – Food Contamination

Scenario 14: Biological Attack – Foreign Animal Disease (Foot and Mouth Disease)

Scenario 15: Cyber Attack

Manmade threats include threats of terrorism, technological accidents, assassinations, kidnappings, hijackings, and cyber attacks (computer-based), and the use of CBRE weapons. High-risk targets include military and civilian government facilities, international airports, large cities, and high-profile landmarks. Terrorists might also target large public gatherings, water and food supplies, utilities, and corporate centers. Further, they are capable of spreading fear by sending explosives or chemical and biological agents through the mail.

This chapter also considers shelter design concepts that relate to the type of shelter being designed and where it may be located. It discusses how shelter use (either single or multiple) may affect the type of shelter selected and the location of that shelter on a particular site. The chapter describes key operations zones in and around a shelter that need to be taken into consideration as a means to provide safe ingress and egress and medical assistance to victims of a manmade event (terrorist attack or technological accident). The decision to enter a shelter is made by the senior management staff based on notification of a credible threat or as a result of an actual disaster. The National Incident Management System (NIMS) and the Catastrophic Incident Supplement (CIS) to the NRP established the procedures to respond to and recover from a CBRE event. Section 4.2 discusses the plan's alerting and notification, and response and recovery processes. The objective of this chapter is to provide a broad vision on how a shelter should be designed to protect against catastrophic events.

The decision to design and construct a shelter can be based on a single factor or on a collection of factors. Single factors are often related to the potential for loss of life or injury (e.g., a hospital that cannot move patients housed in an intensive care unit decides to build a shelter, or shelters, within the hospital; a school decides not to chance fate and constructs a shelter). A collection of factors could include the type of hazard event, probability of event occurrence, severity of the event, probable single and aggregate annual event deaths, shelter costs, and results of computer models that evaluate the benefits and costs of the shelter project.

1.2 POTENTIAL THREATS

Rather than identify a specific threat, this document provides general guidance that will address different types of building construction and the reasonable mitigative measures to provide a secure shelter. However, it is important for building owners and design professionals to understand the potential threats to which buildings may be exposed. This section provides an overview of manmade threats.

The term "threat" is typically used to describe the design criteria for manmade disasters (technological accident) or terrorism. Identifying the threats for manmade threats can be a difficult task. Because they are different from other natural hazards such as earthquakes, floods, and hurricanes, manmade threats are difficult to predict. Many years of historical and quantitative data, and probabilities associated with the cycle, duration, and magnitude of natural hazards exist. The fact that data for manmade threats are scarce and that the magnitude and recurrence of terrorist attacks are almost unpredictable makes the determination of a particular threat for any specific site or building difficult and largely subjective. Such asymmetrical threats do not exclusively target buildings and may employ diversionary tactics to actually direct occupants to a primary attack instrument.

With any manmade threat, it is important to determine who has the intent to cause harm. The aggressors seek publicity for their cause, monetary gain (in some instances), or political gain through their actions. These actions can include injuring or killing people; destroying or damaging facilities, property, equipment, or resources; or stealing equipment, material, or information.

Aggressor tactics run the gamut: moving vehicle bombs; stationary vehicle bombs; bombs delivered by persons (suicide bombers); exterior attacks (thrown objects like rocks, Molotov cocktails, hand grenades, or hand-placed bombs); stand-off weapons attacks (rocket propelled grenades, light antitank weapons, etc.); ballistic attacks (small arms and high power rifles); covert entries (gaining entry by false credentials or circumventing security with or without weapons); mail bombs (delivered to individuals); supply bombs (larger bombs processed through shipping departments); airborne contamination (CBR agents used to contaminate the air supply of a building); and waterborne contamination (CBR agents injected into the water supply). This section focuses on explosive threats, chemical agents, biological warfare agents, and radiological attacks.

1.2.1 Explosive Threats

The explosive threat is particularly insidious, because all of the ingredients required to assemble an improvised explosive device are readily available at a variety of farm and hardware stores. The intensity of the explosive detonation is limited by the expertise of the person assembling the device and the means of delivery. Although the weight of the explosive depends on the means of transportation and delivery, the origin of the threat depends primarily on the access available to the perpetrator. Operational security procedures will define the areas within or around a building at which a device may be located, undetected by the building facilities staff. These security procedures include screening of vehicles, inspection of delivered parcels, and vetting hand carried bags. The extent to which this inspection is carried out will determine the size of an explosive device that may evade detection. Despite the most vigilant attempts, however, it is unrealistic to expect complete success in preventing a small threat from evading detection. Nevertheless, it is unlikely that a large threat may be brought into a building. As a result, a parcel sized device may be introduced into publicly accessible lobbies, garages, loading docks, cafeterias, or retail spaces and it must be assumed that a smaller explosive device may be brought anywhere into the building.

Although operational security measures can drastically limit the size of the explosive device that could be introduced onto a building site, there is no means of limiting the size of the explosive that could be contained within a vehicle traveling on the surrounding streets or roadways.

Explosives weigh approximately 100 pounds per cubic foot and, as a result, the maximum credible threat corresponds to the weight of explosives that can be packaged in a variety of containers or vehicles. The Department of Defense (DoD) developed a chart to help indicate the weight of explosives and deflagrating materials that may reasonably fit within a variety of containers and vehicles (see Table 1-1). The table also indicates the safe evacuation distances for occupants of conventional unreinforced buildings, based on their ability to withstand severe damage or resist collapse. Similarly, Table 1-1 indicates the safe evacuation distance for pedestrians exposed to explosive effects based on the greater of fragment throw distance or glass breakage/falling glass hazard distance. Because a pipe bomb, suicide belt/vest, backpack, and briefcase/suitcase bomb are specifically designed to throw fragments, protection from these devices may require greater safe evacuation distances than an equal weight of explosives transported in a vehicle. Table 1-2 shows safe evacuation distances for liquefied petroleum gas (LPG) threats.

Table 1-1: Safe Evacuation Distances from Explosive Threats

Threat Description		Explosives Mass* (TNT equivalent)	Building Evacuation Distance**	Outdoor Evacuation Distance***
High Explosives (TNT Equivalent)	Pipe Bomb	5 lbs 2.3 kg	70 ft 21 m	850 ft 259 m
	Suicide Belt	10 lbs 4.5 kg	90 ft 27 m	1,080 ft 330 m
	Suicide Vest	20 lbs 9 kg	110 ft 34 m	1,360 ft 415 m
	Briefcase/ Suitcase Bomb	50 lbs 23 kg	150 ft 46 m	1,850 ft 564 m
	Compact Sedan	500 lbs 227 kg	320 ft 98 m	1,500 ft 457 m
	Sedan	1,000 lbs 454 kg	400 ft 122 m	1,750 ft 534 m
	Passenger/ Cargo Van	4,000 lbs 1,814 kg	640 ft 195 m	2,750 ft 838 m
	Small Moving Van/ Delivery Truck	10,000 lbs 4,536 kg	860 ft 263 m	3,750 ft 1,143 m
	Moving Van/ Water Truck	30,000 lbs 13,608 kg	1,240 ft 375 m	6,500 ft 1,982 m
	Semi-trailer	60,000 lbs 27,216 kg	1,570 ft 475 m	7,000 ft 2,134 m

*   Based on the maximum amount of material that could reasonably fit into a container or vehicle. Variations are possible.

**   Governed by the ability of an unreinforced building to withstand severe damage or collapse.

***   Governed by the greater of fragment throw distance or glass breakage/falling glass hazard distance. These distances can be reduced for personnel wearing ballistic protection. Note that the pipe bombs, suicide belts/vests, and briefcase/suitcase bombs are assumed to have a fragmentation characteristic that requires greater stand-off distances than an equal amount of explosives in a vehicle.

Table 1-2: Safe Evacuation Distances from LPG Threats

Threat Description		LPG Mass/Volume	Fireball Diameter*	Safe Distance**
Liquefied Petroleum Gas (LPG - Butane or Propane)	Small LPG Tank	20 lbs/5 gal 9 kg/19 l	40 ft 12 m	160 ft 48 m
	Large LPG Tank	100 lbs/25 gal 45 kg/95 l	69 ft 21 m	276 ft 84 m
	Commercial/ Residential LPG Tank	2,000 lbs/500 gal 907 kg/1,893 l	184 ft 56 m	736 ft 224 m
	Small LPG Truck	8,000 lbs/2,000 gal 3,630 kg/7,570 l	292 ft 89 m	1,168 ft 356 m
	Semi-tanker LPG	40,000 lbs/10,000 gal 18,144 kg/37,850 l	499 ft 152 m	1,996 ft 608 m

*   Assuming efficient mixing of the flammable gas with ambient air.

**   Determined by U.S. firefighting practices wherein safe distances are approximately four times the flame height. Note that an LPG tank filled with high explosives would require a significantly greater stand-off distance than if it were filled with LPG.

The Bureau of Alcohol, Tobacco, Firearms, and Explosives (ATF) report on *Incidents, Casualties and Property Damage* for all states for 2002 lists 553 actual bombing incidents, 32 of which were premature explosions, injuring 80 people, killing 13, and causing over $5 million in damages. Nearly half of the events were against buildings and nearly a quarter were against vehicles.

DESIGN CONSIDERATIONS

Only two domestic terrorist bombings involved the use of large quantities of High Energy explosive materials. (For more information on High Energy explosives, see FEMA 426, *Reference Manual to Mitigate Potential Terrorist Attacks Against Buildings*, Chapter 4.) Although these events represent the largest explosions that have occurred to date, they do not accurately represent the actual domestic explosive threat. The 1995 explosion that collapsed portions of the Murrah Federal Office Building in Oklahoma City contained 4,800 pounds of ammonium nitrate and fuel oil (ANFO) and the 1993 explosion within the parking garage beneath the World Trade Center complex contained 1,200 pounds of urea nitrate.

Every year, approximately 1,000 intentional explosive detonations are reported by the Federal Bureau of Investigation (FBI) Bomb Data Center. As implied by the FBI statistics, the majority of the domestic events contain significantly smaller weights of Low Energy explosives. (For more information on Low Energy explosives, see FEMA 426, Chapter 4.) Figure 1-1 illustrates the breakdown of domestic terrorist events from 1980 to 2001. The vast majority of the 294 terrorist incidents, 55 suspected terrorist incidents, or 133 prevented terrorist incidents, involved explosives and 75 percent of these events occurred in the 1980s. The explosive that was used in the 1996 pipe bomb attack at the Olympics in Atlanta, Georgia, consisted of smokeless powder and was preceded by a warning that was called in 23 minutes before the detonation.

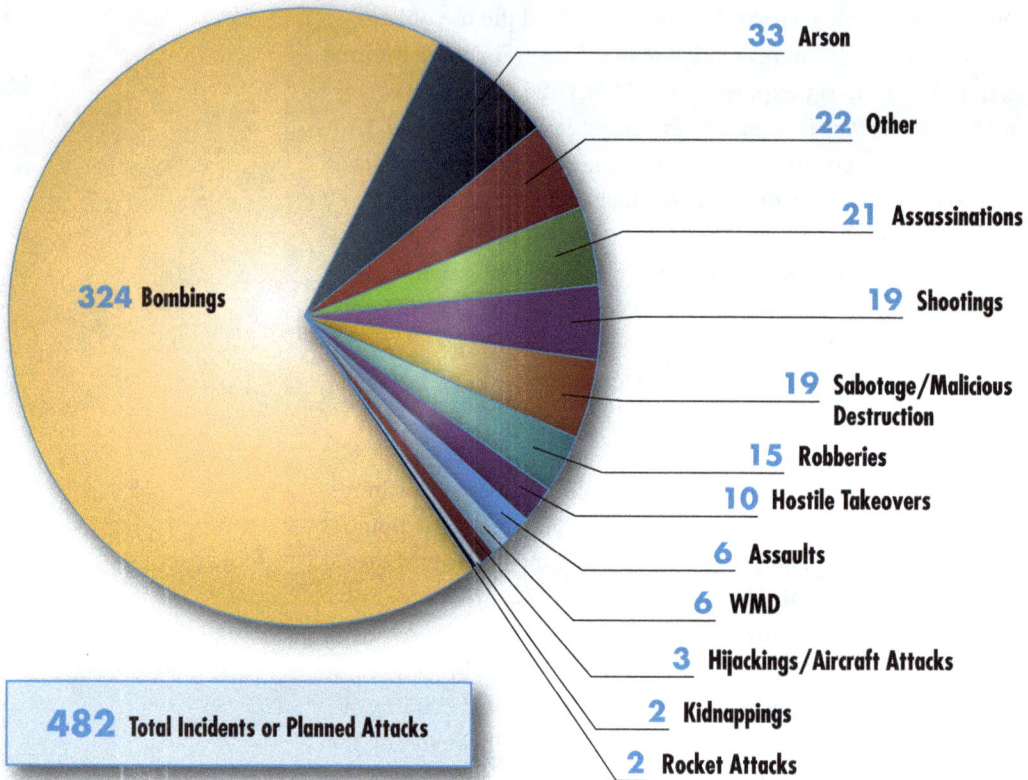

33 Arson	
22 Other	
21 Assassinations	
19 Shootings	
19 Sabotage/Malicious Destruction	
15 Robberies	
10 Hostile Takeovers	
6 Assaults	
6 WMD	
3 Hijackings/Aircraft Attacks	
2 Kidnappings	
2 Rocket Attacks	

324 Bombings

482 Total Incidents or Planned Attacks

Figure 1-1   Terrorism by event 1980 through 2001

SOURCE: FBI TERRORISM 2000/2001 PUBLICATION #308

Although the majority of these explosions targeted residential properties and vehicles, 63 took place in educational facilities, causing a total of $68,500 in property damage. By contrast, other than the attack against the Murrah Federal Office Building, no explosive devices were detonated at a Federal government owned facility, and only nine were detonated at local/state government facilities. Nearly 80 percent of the people known to be involved in bombing incidents were "young offenders," and less than ½ percent of the perpetrators were identified as members of terrorist groups. Vandalism was the motivation in 53 percent of the known intentional and accidental bombing incidents, and the timing of the attacks was fairly uniformly distributed throughout the day.

DESIGN CONSIDERATIONS

Nevertheless, the protective design of structures focuses on the effects of High Energy explosives and relates the different mixtures to an equivalent weight of trinitrotoluene (TNT).

1.2.2 CBR Attacks

Like explosive threats, CBR threats may be delivered externally or internally to the building. External ground-based threats may be released at a stand-off distance from the building or may be delivered directly through an air intake or other opening. Interior threats may be delivered to accessible areas such as the lobby, mailroom, or loading dock, or they may be released into a secure area such as a primary egress route. There may not be an official or obvious warning prior to a CBR event. Although official warnings should always be heeded, the best defense may be to be alert to signs of a release.

There are three potential methods of attacks in terms of CBR:

❍ A large exterior release originating some distance away from the building (includes delivery by aircraft)

❍ A small localized exterior release at an air intake or other opening in the exterior envelope of the building

❍ A small interior release in a publicly accessible area, a major egress route, or other vulnerable area (e.g., elevator lobby, mail room, delivery, receiving and shipping, etc.)

Chapter 4 provides additional guidance on emergency management considerations that may have an impact on siting or design of a shelter.

1.2.2.1 Chemical Agents. Toxic chemical agents can present airborne hazards when dispersed as gases, vapors, or solid or liquid aerosols. Generally, chemical agents produce immediate effects, unlike biological or radiological agents. In most cases, toxic chemical agents can be detected by the senses, although a few are odorless. Their effects occur mainly through inhalation, although they can also cause injury to the eyes and skin.

1.2.2.2 Biological Warfare Agents. Biological warfare agents are organisms or toxins that can kill or incapacitate people and livestock, and destroy crops. The three basic groups of biological agents that would likely be used as weapons are bacteria, viruses, and toxins.

○ **Bacteria.** Bacteria are small free-living organisms that reproduce by simple division and are easy to grow. The diseases they produce often respond to treatment with antibiotics.

○ **Viruses.** Viruses are organisms that require living cells in which to reproduce and are intimately dependent upon the body they infect. The diseases they produce generally do not respond to antibiotics; however, antiviral drugs are sometimes effective.

○ **Toxins.** Toxins are poisonous substances found in, and extracted from, living plants, animals, or microorganisms; some toxins can be produced or altered by chemical means. Some toxins can be treated with specific antitoxins and selected drugs.

Most biological agents are difficult to grow and maintain. Many break down quickly when exposed to sunlight and other environmental factors, while others such as anthrax spores are very long lived. They can be dispersed by spraying them in the air or by infected animals that carry the disease, as well through food and water contamination:

○ **Aerosols.** Biological agents are dispersed into the air, forming a fine mist that may drift for miles. Inhaling the agent may cause disease in people or animals.

○ **Animals.** Some diseases are spread by insects and animals, such as fleas, flies, mosquitoes, and mice. Deliberately spreading diseases through livestock is also referred to as agroterrorism.

DESIGN CONSIDERATIONS

○ **Food and water contamination.** Some pathogenic organisms and toxins may persist in food and water supplies. Most microbes can be killed, and toxins deactivated, by cooking food and boiling water.

Person-to-person spread of a few infectious agents is also possible. Humans have been the source of infection for smallpox, plague, and the Lassa viruses.

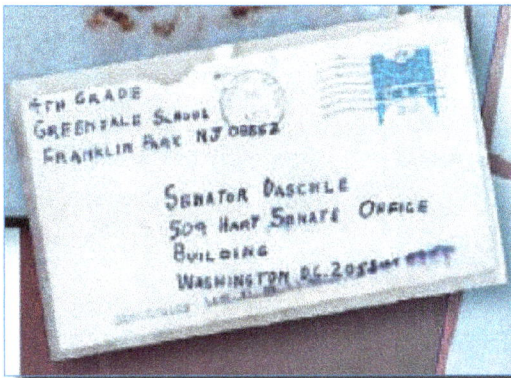

Anthrax spores formulated as a white powder were mailed to individuals in the Federal Government and media in the fall of 2001. Postal sorting machines and the opening of letters dispersed the spores as aerosols. Several deaths resulted. The effect was to disrupt mail service and to cause a widespread fear of handling delivered mail among the public.

Figure 1-2   Sample anthrax letter

SOURCE: FBI TERRORISM 2000/2001 PUBLICATION #308

1.2.2.3 Radiological Attacks. Shelters described in this manual do not address the severe and various effects generated by nuclear events, including blinding light, intense heat (thermal radiation), initial nuclear radiation, blast, fires started by the heat pulse, and secondary fires caused by the destruction. Protection against these severe effects of a nuclear explosion is not considered in this manual.

Terrorist use of a radiological dispersion device (RDD), often called "dirty nuke" or "dirty bomb," is considered far more likely than use of a nuclear device. These radiological weapons are a combination of conventional explosives and radioactive material designed to scatter dangerous and sublethal amounts of radio-active material over a general area. Such radiological weapons appeal to terrorists because they require very little technical knowledge to build and deploy compared to that of a nuclear device. Also, these radioactive materials, used widely in medicine, agriculture, industry, and research, are much more readily available and easy to obtain compared to weapons grade uranium or plutonium. Figure 1-3 shows the number of incidents of radioactive materials smuggling from 1993 to 2003.

Radioactive Smuggling
Smuggling of potential ingredients for a dirty bomb is on the increase, but there are fewer incidents involving fissile material that could be used for making a nuclear bomb.

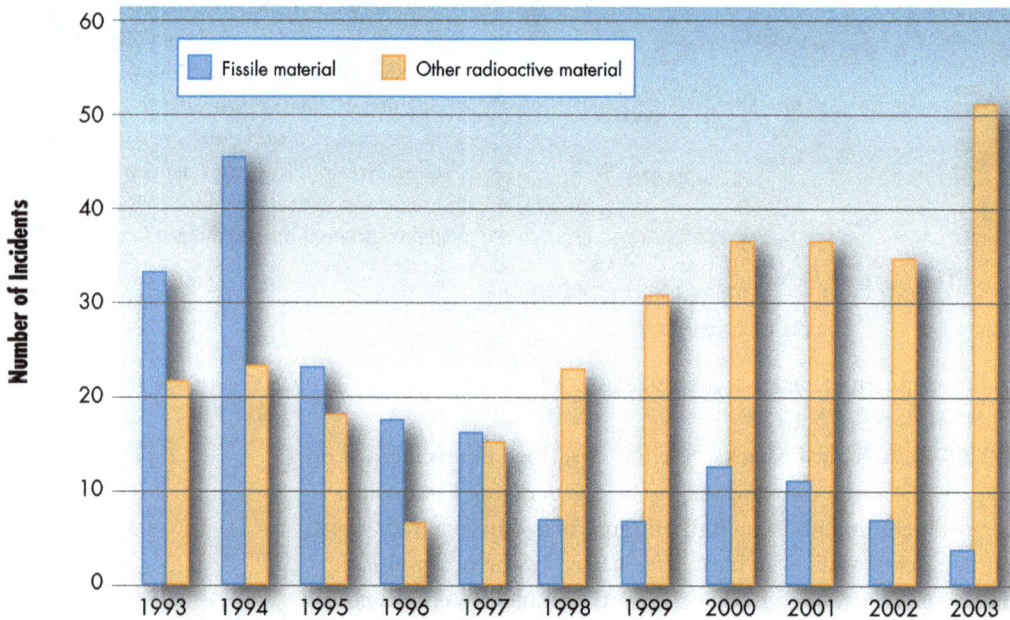

Figure 1-3   Radioactive materials smuggling

SOURCE: INTERNATIONAL ATOMIC ENERGY AGENCY

There is no way of knowing how much warning time there would be before an attack by a terrorist using a radiological weapon. A surprise attack remains a possibility.

1.3 LEVELS OF PROTECTION

Currently, there are only two Federal standards that have been promulgated for Federal facilities that define LOPs for manmade threats: the Interagency Security Committee (ISC) *Design Criteria* and the DoD *Minimum Antiterrorism Standards,* UFC 4-010-01. Both standards address blast primarily through the use of stand-off distance and ensuring walls and glazing blast pressures are strengthened to withstand the blast shock wave. Both standards address CBR agents primarily through the use of filtration, emergency shutdown of mechanical and electrical systems, and mass notification to building occupants.

Until the building, mechanical, electrical, and life safety codes are promulgated for manmade events, the ISC building standards provide a reasonable approach to selecting a level of protection for a shelter for CBR agents.

1.3.1 Blast Levels of Protection

The level of protection in response to blast loading defines the extent of damage and debris that may be sustained in response to the resulting blast pressures and impulses. (For more information on blast pressure impulses, see FEMA 426, Chapter 4.) The levels of protection are generally defined in the terms of performance. Fundamental to the discussion of levels of protection is the notion of repairable damage. Repair is typically assumed to be within days to weeks and the structure requires partial evacuation during repairs. Table 1-3 provides a synopsis of the ISC blast standards.

Table 1-3: Correlation of ISC Levels of Protection and Incident Pressure to Damage and Injury

Level of Protection	Potential Structural Damage	Potential Glazing Hazards
Minimum and Low	The facility or protected space will sustain a high level of damage without progressive collapse. Casualties will occur and assets will be damaged. Building components, including structural members, will require replacement, or the facility may be completely unrepairable, requiring demolition and replacement.	For Minimum Protection, there are no restrictions on the type of glazing used. For Low Protection, there is no requirement to design windows for specific blast pressure loads. However, the use of glazing materials and designs that minimize the risks is encouraged. Glazing cracks and window system fails catastrophically. Fragments enter space, impacting a vertical witness panel at a distance of no more than 3 m (10 ft) from the window at a height greater than 0.6 m (2 ft) above the floor.
Medium	Moderate damage, repairable. The facility or protected space will sustain a significant degree of damage, but the structure will be reusable. Some casualties may occur and assets may be damaged. Building elements other than major structural members may require replacement.	For Medium and High Protection, design up to the specified load as directed by the risk assessment. Window systems design (glazing, frames, anchorage to supporting walls, etc.) on the exterior façade should be balanced to mitigate the hazardous effects of flying glazing following an explosive event. The walls, anchorage, and window framing should fully develop the capacity of the glazing material selected. Glazing cracks. Fragments enter space and land on the floor and impact a vertical witness panel at a distance of no more than 3 m (10 ft) from the window at a height greater than 0.6 m (2 ft) above the floor.
High	Minor damage, repairable. The facility or protected space may globally sustain minor damage with local significant damage possible. Occupants may incur some injury, and assets receive minor damage.	For Medium and High Protection, design up to the specified load as directed by the risk assessment. Window systems design (glazing, frames, anchorage to supporting walls, etc.) on the exterior façade should be balanced to mitigate the hazardous effects of flying glazing following an explosive event. The walls, anchorage, and window framing should fully develop the capacity of the glazing material selected. Glazing cracks. Fragments enter space and land on the floor no farther than 3 m (10 ft) from the window.

DESIGN CONSIDERATIONS

1.3.2 CBR Levels of Protection

Protection against airborne chemical, biological, and radiological (CBR) agents or contaminants is typically achieved by using particulate and adsorption filters, and personal protective equipment (PPE). Many different types of filters are available for CBR releases. Filter efficiency (e.g., how well the filter captures the toxic material) varies based on the filter type (e.g., activated or impregnated charcoal) and the specific toxic material. No single filter can protect against all CBR materials; therefore, it is important to verify which CBR materials a filter protects against.

There are three levels of protection that range from filtration with pressurization (Class 1), filtration with little or no pressurization (Class 2), and passive protection (Class 3). Class 1 protection is for a large-scale release over an extended period of time and would apply to mission essential government and commercial buildings that must remain operational 24 hours a day/7 days a week. Class 2 protection is for a terrorist attack or technological accident with little or no warning and is characterized as a short duration small scale release. Class 3 is typically applicable to an industrial accident that results in a short duration release. These three levels of protection are discussed in greater detail in Chapter 3. Table 1-4 provides a synopsis of the ISC CBR protection standards.

> The CBR levels of protection included in this section are consistent with the Department of Homeland Security (DHS) Working Group on Radiological Dispersal Device Preparedness and the Health Physics Society's (HPS's) Scientific and Public Issues Committee reports:
>
> "Sheltering is 10-80% effective in reducing dose depending upon the duration of exposure, building design and ventilation. If there is a passing plume of radioactivity, sheltering may be preferable to evacuation. When sheltering, ventilation should be turned off to reduce influx of outside air. Sheltering may not be appropriate if doses are projected to be very high or long in duration."
>
> "Sheltering is likely to be more protective than evacuation in responding to a radiological terrorist event. Therefore, the HPS recommends that sheltering be the preferred protective action. The Protective Action Guidance (PAG) for sheltering is the same as the existing evacuation PAG, i.e., 10 mSv (1 rem), with the minimum level for initiation being the same as the existing PAG, i.e., 1 mSv (100 mrem)."

Table 1-4: ISC CBR Levels of Protection

Level of Protection	For Biological/ Radiological Contaminants	For Chemical/ Radiological	Additional Considerations	Class
Low	Use minimum efficiency reporting value (MERV) 13 filter or functional equivalent.	None	None	3
Medium	Use high-efficiency particulate air (HEPA) filter or functional equivalent.	Use gas absorber for outside air.	Design for future detection technology Stairway pressurization system should maintain positive pressure in stairways for occupant refuge, safe evacuation, and access by firefighters. The entry of smoke and hazardous gases into stairways must be minimized. Locate utility systems at least 15 m (50 ft) from loading docks, front entrances, and parking areas.	2
High	Use HEPA filter or functional equivalent.	Use gas absorber for outside air and return air.	Design for future detection technology Stairway pressurization system should maintain positive pressure in stairways for occupant refuge, safe evacuation, and access by firefighters. The entry of smoke and hazardous gases into stairways must be minimized. Locate utility systems at least 15 m (50 ft) from loading docks, front entrances, and parking areas.	1, 2

DESIGN CONSIDERATIONS

1.4 SHELTER TYPES

A CBRE shelter can be designed as a standalone or internal shelter to be used solely as a shelter or to have multiple purposes, uses, or occupancies. This section provides a series of definitions that can be useful when deciding to build a new shelter or upgrade an existing shelter.

1.4.1   Standalone Shelters

A standalone shelter is considered a separate building (i.e., not within or attached to any other building) that is designed and constructed to withstand the range of natural and manmade hazards. This type of shelter has the following characteristics:

○ It may be sited away from potential debris hazards.

○ It will be structurally and mechanically separate from any building and therefore not vulnerable to being weakened if part of an adjacent structure collapses or if a CBRE event occurs in the adjacent building.

○ It does not need to be integrated into an existing building design.

A shelter for CBRE protection may be as simple as an interior residential room to the traditional public shelter able to support several hundred people. The number of persons taking refuge in a shelter will typically be more than 12 and could be up to several hundred or more.

1.4.2   Internal Shelters

An internal shelter is a specially designed and constructed room or area within or attached to a larger building that is designed and constructed to be structurally independent of the larger building and to withstand the range of natural and manmade hazards. It shows the following characteristics:

○ It is partially shielded by the surrounding building and may not experience the full force of the blast. (Note that any protection provided by the surrounding building should not be considered in the shelter design.)

○ It is designed to be within a new building and may be located in an area of the building that the building occupants can reach quickly, easily, and without having to go outside, such as a data center, conference room, gymnasium, or cafeteria.

○ It may reduce the shelter cost because it is typically part of a planned renovation or building project.

1.4.3   Shelter Categories

A standalone or internal shelter may serve as a shelter only, or it may have multiple uses (e.g., a multi-use shelter at a school could also function as a classroom, lunchroom, or laboratory; a multi-use shelter intended to serve a manufactured housing community or single-family-home subdivision could also function as a community center). The decision to design and construct a single-use or a multi-use shelter will likely be made by the prospective client or the owner of the shelter. To help the designer respond to non-engineering and non-architectural needs of shelter owners, this section discusses different shelter categories and usages. Table 1-5 provides a summary of the commercial shelter categories.

DESIGN CONSIDERATIONS

Table 1-5: Commercial Shelter Categories

Shelter Considerations	In-Ground	Single-Use	Multi-Use	Community
Level of Protection	Blast – Medium CBR – Class 3	Blast – Low CBR – Class 3	All	All
Expected Capacity	1-100	1-10	1-100	100-1,000
Location	Basement or sub-basement area without windows and semi-hardened walls and ceiling	Interior space without windows and semi-hardened walls and ceiling	Conference Room Data Center Bathroom Stairwell Elevator Core	School Church Mall Government Building
Special Considerations	Difficult to site/build in high water table and rocky areas	Annual or semi-annual inspection and rotation of supplies	May need multiple areas in large buildings and commercial office space; plan and exercises to prevent overcrowding	Plan for multi-lingual, elderly, non-ambulatory, and special needs populations • Life Safety NFPA 101 and 5000 guidance • ADA compliance

NFP = National Fire Protection Association
ADA = Americans with Disabilities Act

- **In-ground shelters.** The in-ground shelters referred to in this manual are built below ground inside a building and therefore can be entered directly from within the building. Other types of in-ground shelters are available that are designed to be installed outside a building and entering one of these exterior in-ground shelters would require leaving the building.

○ **Single-use shelters.** Single-use shelters are used only in the event of a hazard event. One advantage of single-use shelters is a potentially simplified design that may be readily accepted by the authority having local jurisdiction. These shelters typically have simplified electrical and mechanical systems because they are not required to provide normal daily accommodations for people. Single-use shelters are always ready for occupants and will not be cluttered with furnishings and storage items, which is a concern with multi-use shelters. Simplified, single-use shelters may have a lower total cost of construction than multi-use shelters.

The cost of building a single-use shelter is much higher than the additional cost of including shelter protection in a multi-use room. Existing maintenance plans will usually consider multi-use rooms, but single-use shelters can be expected to require an additional annual maintenance cost.

○ **Multi-use shelters.** The ability to use a shelter for more than one purpose often makes a multi-use standalone or internal shelter appealing to a shelter owner or operator. Multi-use shelters also allow immediate return on investment for owners/operators; the shelter space is used for daily business when the shelter is not being used during a hazard event. Hospitals, assisted living facilities, and special needs centers would benefit from multi-use internal shelters, such as hardened intensive care units or surgical suites. Internal multi-use shelters in these types of facilities allow optimization of space while providing near-absolute protection with easy access for non-ambulatory persons. In new buildings being designed and constructed, recent FEMA-sponsored projects have indicated that the construction cost of hardening a small area or room in a building is 10 to 25 percent higher than the construction cost for a non-hardened version of the same area or room.

○ **Community shelters at neighborhoods and or public facilities**. Community shelters are intended to provide protection for the residents of neighborhoods and are typically located at schools and other similar institutions; they are identified, categorized, and labeled by the American Red Cross (see ARC 4496).

1.5 SITING

One of the most important elements in designing a shelter is its location or siting. In inspecting areas of existing buildings that are used as shelter areas, research has found that owners may overlook the safest area of a building, while the safety of a hallway or other shelter areas may be overestimated. Evaluating shelter areas in an existing building or determining the best areas for new ones is invaluable for saving lives when a disaster strikes.

The location of a shelter on a building site is an important part of the design process for shelters. The shelter location on the site and capacity should consider how many occupants work in the building, as well as how many non-occupants may take refuge in the nearest shelter available. At the site and building level, the shelter location analysis should include evaluation of potential CBRE effects.

When deciding to build a shelter, a preliminary evaluation may be performed by a design professional or by a potential shelter owner, property owner, emergency manager, building maintenance person, or other interested party provided he or she has a basic knowledge of building sciences and can understand building design plans and specifications. **Although the threat of damage from CBRE events may be the predominant focus of the evaluation, additional threats may exist from tornado, hurricane, flood, and seismic events; therefore, the evaluation should assess the threat at the site**. Prior to the design and construction of a shelter, a design professional should perform a more thorough assessment in order to confirm or, as necessary, modify the findings of a preliminary assessment.

An entire building or a section of a building may be designated as a potential shelter area. To perform an assessment of an existing structure or a new structure to be used as a shelter, the building owner or designers may use the Building Vulnerability Assessment Checklist included in FEMA 426, *Reference Manual to Mitigate Potential Terrorist Attacks Against Buildings;* FEMA 452, *A How-To Guide to Mitigate Potential Terrorist Attacks Against Buildings* for the assessment of CBRE events; and FEMA 433, *Using HAZUS-MH for Risk Assessment* for the assessment of major natural hazards.

If an existing building is selected for use as a shelter, the Building Vulnerability Assessment Checklist will help the user identify major vulnerabilities and/or the best shelter areas within the building to place the shelter. The checklist consists of questions pertaining to structural, nonstructural, and mechanical characteristics of the area being considered. The questions are designed to identify structural, nonstructural, and mechanical vulnerabilities to CBRE hazards based on typical failure mechanisms. Structural, nonstructural, and mechanical deficiencies may be remedied with retrofit designs; however, depending on the type and degree of deficiency, the evaluation may indicate that the existing structure is unsuitable for use as a shelter area. A detailed analysis should consider if a portion of a particular building can be used as shelter or whether that portion is structurally independent of the rest of the building. It should also determine if the location is easily accessible, contains the required square footage, and has good ingress and egress elements.

The shelter should be located such that all persons designated to take refuge may reach the shelter with minimal travel time. Shelters located at one end of a building or one end of a community, office complex, or school may be difficult for some users at a site to reach in a timely fashion. Routes to the shelter should be easily accessible and well marked. Exit routes from the shelter should be in a direction away from the threat. Hazard signs should be located following Crime Prevention Through Environmental Design (CPTED) principles of natural access control, natural surveillance, and territoriality and illustrated in Figure 1-4.

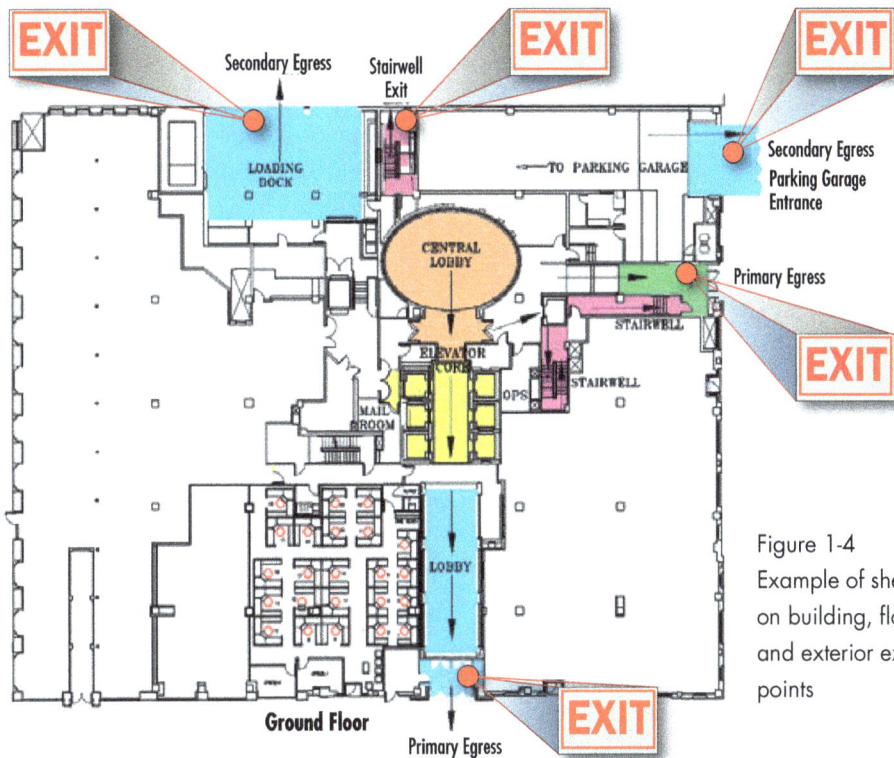

Figure 1-4
Example of shelter marking on building, floor plan, and exterior exits to rally points

○ **Natural access control (controls access).** Guides people entering and leaving a space through the placement of entrances, exits, fences, landscaping, and lighting. Access control can decrease opportunities for terrorist activity by denying access to potential targets and creating a perception of risk for would-be terrorists.

○ **Natural surveillance (increases visibility).** The placement of physical features, activities, and people in a way that maximizes visibility. A potential criminal is less likely to attempt an act of terrorism if he or she is at risk of being observed. At the same time, we are likely to feel safer when we can see and be seen.

○ **Territoriality (promotes a sense of ownership).** The use of physical attributes that express ownership such as fences, signage, landscaping, lighting, pavement designs, etc. Defined property lines and clear distinctions between private and public spaces are examples of the application of territoriality. Territoriality can be seen in gateways into a community or neighborhood.

Shelters should also be located outside areas known to be flood-prone, including areas within the 100-year floodplain. Shelters in flood-prone areas will be susceptible to damage from hydrostatic and hydrodynamic forces associated with rising flood waters. Damage may also be caused by debris floating in the water. Most importantly, flooding of occupied shelters may well result in injuries or deaths. Furthermore, shelters located in flood-prone areas, but properly elevated above the 100-year flood elevation, could become isolated if access routes were flooded. As a result, shelter occupants could be injured and no emergency services would be available.

Where possible, the shelter should be located away from large objects and multi-story buildings. Light towers, antennas, satellite dishes, and roof-mounted mechanical equipment may be toppled or become airborne during blast, hurricane, tornado, or earthquake events. Multi-story buildings adjacent to a shelter may be damaged or may fail structurally due to natural or manmade hazards. When these types of objects or structures fail, they may damage the shelter by collapsing onto it or impacting it. The impact forces associated with these objects are well outside the design parameters of any building code.

There are several possible locations in a building or a house for a shelter. Perhaps the most convenient and safest is below ground level, in a basement. If the building or house does not have a basement, an in-ground shelter can be installed beneath a concrete slab-on-grade foundation or a concrete garage floor (typically would be used as a single-use shelter). Basement

DESIGN CONSIDERATIONS

shelters and in-ground shelters provide the greatest degree of protection against missiles and falling debris.

Another alternative shelter location is an interior room on the first floor of a building or house. Closets, bathrooms, and small storage rooms offer the advantage of having a function other than providing occasional storm protection. Typically, these rooms have only one door and no windows, which make them well-suited for conversion to a shelter. Bathrooms have the added advantage of a water supply and toilet.

Regardless of where in a building or house a shelter is built, the walls and ceiling of the shelter must be built so that they will protect the occupants from missiles and falling debris, and so that they will remain standing if the building or house is severely damaged by extreme winds. If sections of the building or house walls are used as shelter walls, those sections must be separated from the structure of the building or house. This is true regardless of whether interior or exterior walls of the building or house are used as shelter walls.

Typical floor plans of possible locations for shelters in a home are highlighted in yellow in Figures 1-5 and 1-6. These are not floor plans developed specifically for houses with shelters, but they show how shelters can be added without changes to the layout of rooms.

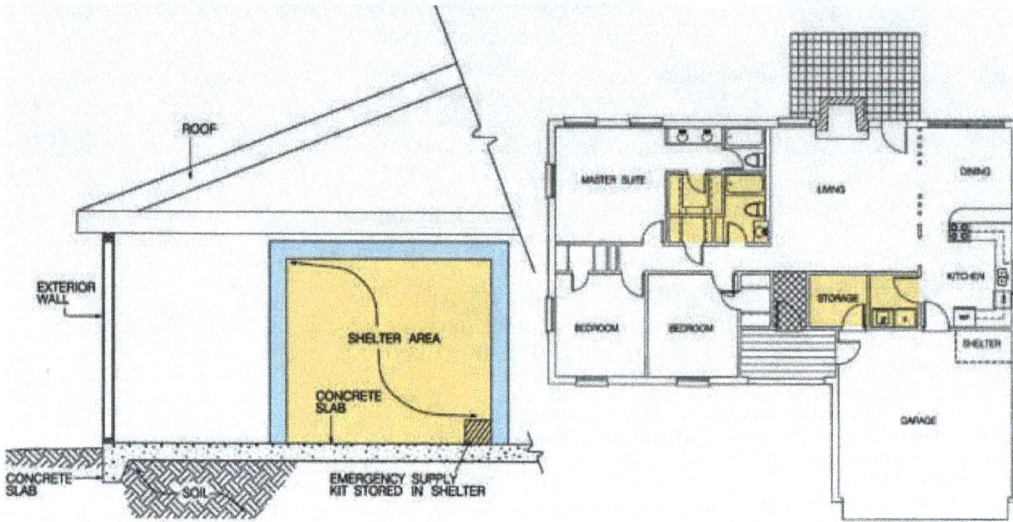

Figure 1-5   Examples of internal shelter locations in a residential slab on grade foundation
SOURCE: FEMA 320

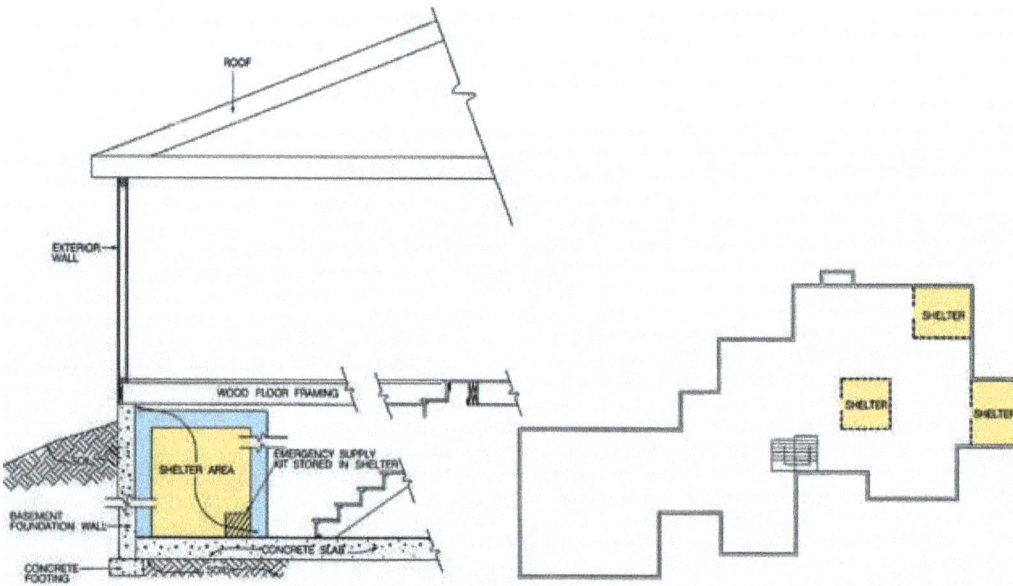

Figure 1-6   Examples of internal shelter locations in a residential basement
SOURCE: FEMA 320

DESIGN CONSIDERATIONS

Figures 1-7 through 1-9 show examples of internal shelter locations in a commercial basement, concourse, and underground parking garage; a retail/commercial multi-story building using a parking garage, conference rooms, data centers, stairwells, and elevator core areas; and a school/church facility, respectively.

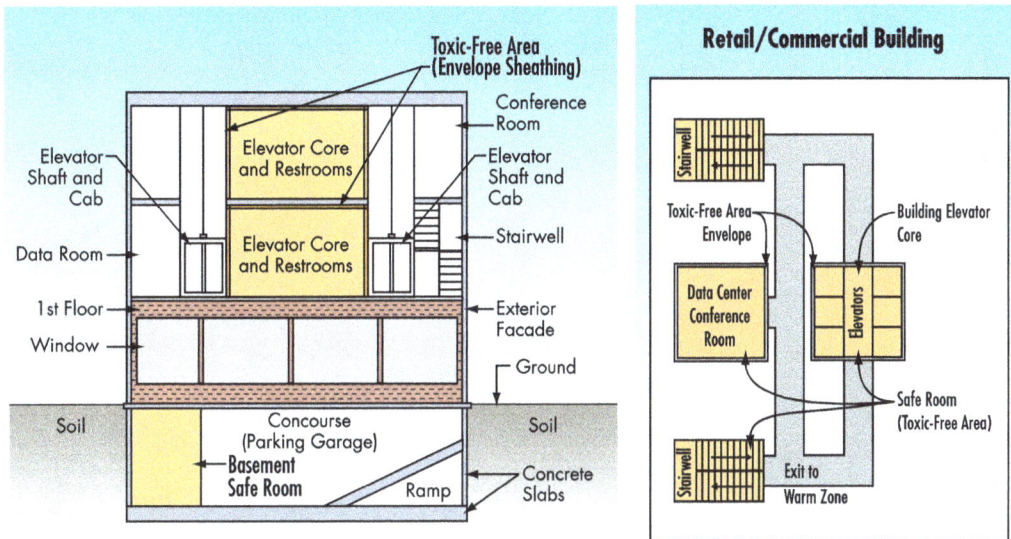

Figure 1-7   Examples of internal shelter locations in a commercial building

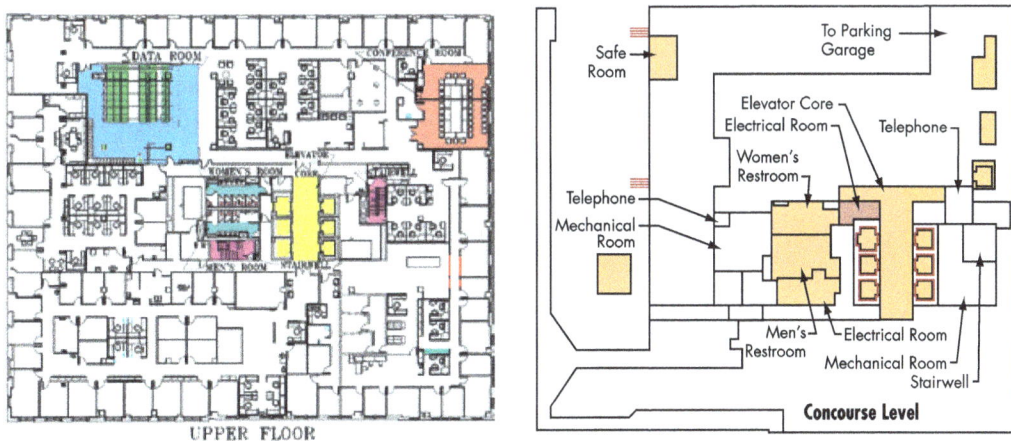

Figure 1-8   Examples of internal shelter locations in a retail/commercial multi-story building using parking garage, conference rooms, data centers, stairwells, and elevator core areas

Figure 1-9
Examples of internal shelter
locations in a school/church
facility

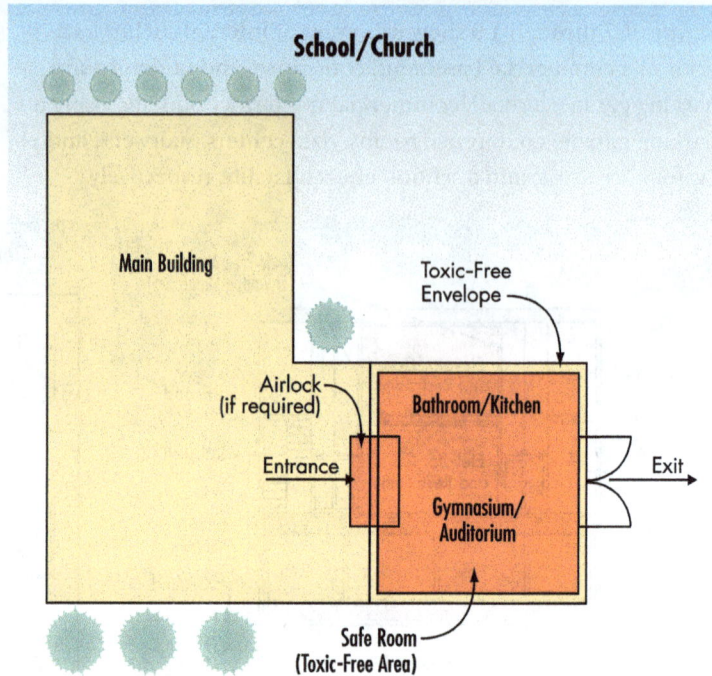

Figure 1-9 Examples of internal shelter locations in a school/church facility

Currently, standalone shelters are relatively rare and most remaining shelters are remnants of the Cold War era that were designed for nuclear weapons protection as "fallout shelters." These shelters were called "dedicated shelters" to make a clear differentiation from dual use shelters (normal facilities in the community that had enhanced radiation protection). Dedicated shelters were built with very high levels of protection and did not have peace time functional compromises. Siting of standalone shelters for nuclear protection has typically been underground, as tunnels, caves, or buried structures. The mass of the geological materials absorbed the blast energy and provided radiation shielding.

Many of the siting and design principles developed by the Office of Civil Defense in FEMA TR-29, *Architect & Engineer Activities in Shelter Development;* FEMA RR-7, *Civil Defense Shelters A State of the Art Assessment 1986;* and FEMA TR-87, *Standards for Fallout Shelters* are still applicable.

For a standalone shelter, many sites will be constrained or site limited for underground, and an aboveground structure may be the only feasible alternative. For these sites, the siting considerations include:

○ Outside the floodplain

○ Separation distance between buildings and structures to prevent progressive collapse or impact from collapsing elements

○ Separation from major transportation features (road, rail)

○ Access to redundant power and communications capabilities

1.6 OCCUPANCY DURATION, TOXIC-FREE AREA (TFA) FLOOR SPACE, AND VENTILATION REQUIREMENTS

Occupancy duration (also known as button-up time) is the length of time that people will be in the shelter with the doors closed and in the protected environment. This period of time is determined by the building owner or local authorities and can range from several hours to several days. For off-site industrial accidents, the occupancy duration is usually less than 24 hours; occupancy durations longer than 24 hours are generally restricted to wartime. Occupancy duration stops when the doors to the shelter are opened. It influences the floor area requirements and the amount of consumable and waste storage. Generally, occupancy duration will not significantly affect the performance of the collective protection system.

> CBR Collective Protection Shelter Basics
>
> ○ Occupancy Duration
>
> ○ Toxic-free Area (TFA) Floor Space
>
> ○ Ventilation Requirements

a. *Less Than 24 Hours.* An occupancy duration of less than 24 hours does not require sleeping areas. The occupant load will generally be a net 1.86 m²/person (20 square feet/person), depending upon the classification of occupancy. The classification of occupancy, as stated in NFPA 101, may require a higher or lower occupant loading depending upon the building classification. The occupant loading will be coordinated with the authority having jurisdiction.

b. *More Than 24 Hours.* An occupancy duration greater than 24 hours requires sleeping areas. The minimum floor area, with the use of single size beds, is approximately 5.6 m^2/person (60 square feet/person). With the use of bunked beds, the minimum floor area is approximately 2.8 m^2/person (30 square feet/person).

The total required TFA floor space is determined from the occupancy duration, the number of people sheltered, and the required floor area per person. Generally, large open areas such as common areas, multi-purpose areas, gymnasiums, etc., provide the most efficient floor area for protecting a large number of personnel. The TFA envelope should include bathroom facilities and, if possible, kitchen facilities.

Although the planned response to CBR events may be to temporarily deactivate the ventilation systems, both single- and multi-use shelters must include ventilation systems capable of providing the minimum number of air changes required by the building code for the shelter's occupancy classification. This will provide a flushing capability once the CBR hazard has passed and facilitate use of the shelter for non-CBR events. For single-use shelters, 15 cubic feet per person per minute is the minimum air exchange recommended; this recommendation is based on guidance outlined in the International Mechanical Code (IMC). For multi-use shelters, the design of mechanical ventilation systems is recommended to accommodate the air exchange requirements for the occupancy classification of the normal use of the shelter area. Although the ventilation system may be overwhelmed in a rare event when the area is used as a shelter, air exchange will still take place. The designer should still confirm with the local building official that the ventilation system may be designed for the normal-use occupancy. In the event the community where the shelter is to be located has not adopted a model building and/or mechanical code, the requirements of the most recent edition of the International Building Code (IBC) are recommended.

1.7 HUMAN FACTORS CRITERIA

Human factors criteria for the natural and manmade hazard shelters build on existing guidance provided in this chapter and in FEMA 320 and 361. Although existing documents do not address all the human factors involved in the design of CBRE shelters, they provide the basis for the criteria summarized in this chapter. These criteria are detailed in the following sections.

1.7.1 Square Footage/Occupancy Requirements

The duration of occupancy of a shelter will vary, depending on the intended event for which the shelter has been designed. Occupancy duration is an important factor that influences many aspects of the design process.

The recommended minimums are 5 square feet per person for tornado shelters and 10 square feet per person for hurricane shelters. The shelter designer should be aware of the occupancy requirements of the building code governing the construction of the shelter. The occupancy loads in the building codes have historically been developed for life safety considerations. Most building codes will require the maximum occupancy of the shelter area to be clearly posted. Multi-use occupancy classifications are provided in the IBC; NFPA 101, *Life Safety Code;* NFPA 5000, *Building Construction and Safety Code;* and state and local building codes. Conflicts may arise between the code-specified occupancy classifications for normal use and the occupancy needed for sheltering. For example, according to the IBC and NFPA 101 and 5000, the occupancy classification for educational use is 20 square feet per person; however, the recommendation for a tornado shelter is 5 square feet per person. Without proper signage and posted occupancy requirements, using an area in a school as a shelter can create a potential conflict regarding the allowed number of persons in the shelter. If both the normal maximum occupancy and the shelter maximum occupancy are posted, and the shelter occupancy is not based on a minimum less than the recommended 5 square feet per person, the shelter design should be acceptable to the building official. The IBC, NFPA 101 and 5000, and the model building codes all have provisions that allow occupancies as concentrated as 5 square feet per person. The American Society of Heating, Refrigeration, and Air-Conditioning Engineers (ASHRAE) recommends that a minimum head room of 6.5 feet and a minimum of 65 cubic feet of net volume be provided per shelter occupant. Net volume shall be determined using the net area calculated for the space.

ASHRAE Ventilation Standard 62-1981, *Ventilation for Smoking-Permitted Areas* defines minimum outdoor air supply rates for various types of occupancy. These rates have been arrived at through a consensus of experts working in the field. A minimum rate of 5 cfm per person for sedentary activity and normal diet holds the carbon dioxide (CO_2) level in a space at 0.25 percent under steady state conditions. Although normal healthy people tolerate 0.5 percent CO_2 without undesirable symptoms and nuclear submarines sometimes operate with 1 percent CO_2 in the atmosphere, a level of 0.25 percent provides a safety factor for increased activity, unusual occupancy load, or reduced ventilation. The ASHRAE Handbook *1982 Applications Environmental Control for Survival* states that carbon dioxide concentration should not exceed 3 percent by volume and preferably should be maintained below 0.5 percent. For a sedentary man, 3 cfm per person of fresh air would maintain a CO_2 concentration of 0.5 percent.

1.7.1.1 Tornado or Short-term Shelter Square Footage Recommendations. Historical data indicate that tornado shelters will typically have a maximum occupancy time of 2 hours. Because the occupancy time is so short, many items that are needed for the comfort of occupants for longer durations (in hurricane shelters) are not recommended for a tornado shelter. FEMA 361, Section 8.2 recommends a minimum of 5 square feet per person for tornado shelters. However, other circumstances and human factors may require the shelter to accommodate persons who require more than 5 square feet. Square footage recommendations for persons with special needs are presented below; these recommendations are the same as those provided in the FEMA *1999 National Performance Criteria for Tornado Shelters*:

○ 5 square feet per person adults standing

○ 6 square feet per person adults seated

○ 5 square feet per person children (under the age of 10)

○ 10 square feet per person wheelchair users

○ 30 square feet per person bedridden persons

1.7.1.2 Hurricane or Long-term Shelter Square Footage Recommendations. Historical data indicate that hurricane shelters will typically have a maximum occupancy time of 36 hours. For this reason, the occupants of a hurricane shelter need more space and comforts than the occupants of a tornado shelter. FEMA 361, Section 8.2 recommends a minimum of 10 square feet per person for hurricane shelters (for a hurricane event only; an event expected to last less than 36 hours). The American Red Cross 4496 publication recommends the following minimum floor areas (Note: the ARC square footage criteria are based on long-term use of the shelter [i.e., use of the shelter both as a refuge area during the event and as a recovery center after the event]):

○ 20 square feet per person for a short-term stay (i.e., a few days)

○ 40 square feet per person for a long-term stay (i.e., days to weeks)

Again, the designer should be aware that there can be conflicts between the occupancy rating for the intended normal use of the shelter and the occupancy required for sheltering. This occupancy conflict can directly affect exit (egress) requirements for the shelter.

1.7.2 Distance/Travel Time and Accessibility

The shelter designer should consider the time required for all occupants of a building or facility to reach the shelter. The National Weather Service (NWS) has made great strides in predicting tornadoes and hurricanes and providing warnings that allow time to seek shelter; it has now expanded the service to include all hazards.

As part of the NIMS, for tornadoes, the time span is often short between the NWS warning and the onset of the tornado. Figure 1-10 shows a sample NWS current watches, warnings, statements, and advisories summary. This manual recommends that a tornado shelter be designed and located in such a way that the following access criteria are met: all potential users of the shelter should be able to reach it within 5 minutes, and the shelter doors should be secured within 10 minutes. For hurricane shelters, these restrictions do not apply, because warnings are issued much earlier, allowing more time for preparation. A CBRE event may have warning such as the Irish Republican Army gave to London police and residents, or no warning as happened with the events of 9/11, and anthrax and sarin releases in October 2001 and the Tokyo subway, respectively.

Figure 1-10
National Weather Service
forecast and warnings
SOURCE: NWS

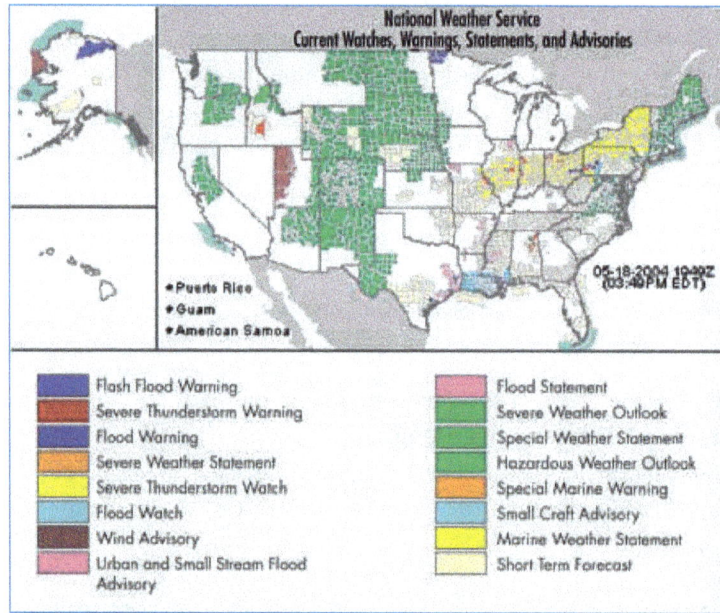

Travel time may be especially important when shelter users have disabilities that impair their mobility. Those with special needs may require assistance from others to reach the shelter; wheelchair users may require a particular route that accommodates the wheelchair. The designer must consider these factors in order to

DESIGN CONSIDERATIONS

provide the shortest possible access time and most accessible route for all potential shelter occupants.

Access is an important element of shelter design. If obstructions exist along the travel route, or if the shelter is cluttered with non-essential equipment and storage items, access to the shelter will be impeded. It is essential that the path remain unencumbered to allow orderly access to the shelter. Hindering access in any way can lead to chaos and panic. For example, at a community shelter built to serve a residential neighborhood, parking at the shelter site may complicate access to the shelter; at a non-residential shelter, such as at a manufacturing plant, mechanical equipment can impede access.

Unstable or poorly secured building elements could potentially block access if a collapse occurs that creates debris piles along the access route or at entrances. A likely scenario is an overhead canopy or large overhang that lacks the capacity to withstand blast effects collapses over the entranceway. The inclusion of these elements should be seriously considered when designing access points in shelters.

1.7.3   Americans with Disabilities Act (ADA)

The needs of persons with disabilities requiring shelter space should be considered. The appropriate access for persons with disabilities must be provided in accordance with all Federal, state, and local ADA requirements and ordinances. If the minimum requirements dictate only one ADA-compliant access point for the shelter, the design professional should consider providing a second ADA-compliant access point for use in the event that the primary access point is blocked or inoperable. Additional guidance for compliance with the ADA can be found in many privately produced publications.

The design professional can ensure that the operations plan developed for the shelter adheres to requirements of the ADA by assisting the owner/operator of the shelter in the development of the plan. All shelters should be managed with an operations and maintenance plan. Developing a sound operations plan is extremely important if compliance with ADA at the shelter site requires the use of lifts, elevators, ramps, or other considerations for shelters that are not directly accessible to non-ambulatory persons.

1.7.4   Special Needs

The use of the shelter also needs to be considered in the design. Occupancy classifications, life safety code, and ADA requirements may dictate the design of such elements as door opening sizes and number of doors, but use of the shelter by hospitals, nursing homes, assisted living facilities, and other special needs groups may affect access requirements to the shelter. For example, basic requirements are outlined in the IBC and NFPA 101 and 5000 regarding the provision of uninterruptible power supplies for life support equipment (e.g., oxygen) for patients in hospitals and other health care facilities. NFPA 99, *Standard for Healthcare Facilities*, specifies details on subjects such as the type, class, and duration of power supplies necessary for critical life support equipment. In addition, it also details the design, arrangement, and configuration of medical gas piping systems, alarms, and networks.

In addition, strict requirements concerning issues such as egress, emergency lighting, and detection-alarm-communication systems are presented in Chapter 10 of the IBC and in NFPA 101, 2006 Edition, Chapters 18 and 19, for health care occupancies. The egress requirements for travel distances, door widths, and locking devices on doors for health care occupancies are more restrictive than those for an assembly occupancy classification in non-health care facilities based on the model building codes for non-health care facilities. Additional requirements also exist for health care facilities that address automatic fire doors, maximum allowable room sizes, and maximum allowable distances to egress points. The combination of all these requirements could lead to the construction of multiple small shelters in a health care facility rather than one large shelter.

1.8 OTHER DESIGN CONSIDERATIONS

Emergency lighting and power, as well as a backup power source, need to be included in the design of multi-use shelters. Route marking and wayfinding, and signage also should be included.

1.8.1 Lighting

For the regular (i.e., non-shelter) use of multi-use shelters, lighting, including emergency lighting for assembly occupancies, is required by all model building codes. Emergency lighting is recommended for community shelters. A backup power source for lighting is essential during a disaster because the main power source is often disrupted. A battery-powered system is recommended as the backup source because it can be located, and fully protected, within the shelter. Flashlights stored in cabinets are useful as secondary lighting provisions, but should not be used as the primary backup lighting system.

A reliable lighting system will help calm shelter occupants during a disaster. Failing to provide proper illumination in a shelter may make it difficult for shelter owners/operators to minimize the agitation and stress of the shelter occupants during the event. If the backup power supply for the lighting system is not contained within the shelter, it should be protected with a structure designed to the same criteria as the shelter itself. Natural lighting provided by windows and doors is often a local design requirement, but is not required by the IBC for assembly occupancies. The 2003 edition of the IBC and the 2006 edition of NFPA 5000 has additional guidance on egress, lighting, and markings.

1.8.2 Emergency Power

Shelters will have different emergency (backup) power needs based upon the length of time that people will stay in the shelters (i.e., shorter duration for tornadoes and longer duration for hurricanes). In addition to the essential requirements that must be provided in the design of the shelter, comfort and convenience should be addressed.

For tornado shelters, the most critical use of emergency power is for lighting. Emergency power may also be required in order to meet the ventilation requirements described in Chapter 3 and Section 1.7.1. The user of the shelter should set this requirement for special needs facilities, but most tornado shelters would not require additional emergency power.

For hurricane shelters, emergency power may be required for both lighting and ventilation. This is particularly important for shelters in hospitals and other special needs facilities. Therefore, a backup generator is recommended. Any generator relied on for emergency power should be protected with an enclosure designed to the same criteria as the shelter.

As illustrated in the previous sections, the manmade hazards shelter design criteria require an adjustment to the traditional design process for natural hazard shelters. The shelter location, operation, and life-cycle costs are now significantly coupled to the community, first responders, and government plans and procedures for mass casualty response and recovery; Federal and local laws for criminal investigation; and the unique site and building design parameters and level of protection that is desired.

1.8.3   Route Marking and Wayfinding

Route marking or wayfinding in an emergency situation such as total darkness has historically relied on fire exit lighting. A new technology that is being adopted by many cities is photoluminescent exit path marking. These photoluminescent self-adhesive signs and tapes are very visible during the day and will glow for up to 8 hours after the light source is removed. These signs have durable, permanent, and renewable fluorescence. Figure 1-11 shows sample signs.

Figure 1-11   Photoluminescent signs, stair treads, and route marking

1.8.4   Signage

The signs should be illuminated, luminescent, and obvious. Key elements of signage include the following.

1.8.4.1 Community and Parking Signage. It is very important that shelter occupants can reach the shelter quickly and without chaos. Parking is often a problem at community shelters; therefore, a Community Shelter Operations Plan should instruct occupants to proceed to a shelter on foot if time permits. Main pathways should be determined and laid out for the community. Pathways should be marked to direct users to the shelter. Finally, the exterior of the shelter should have a sign that clearly identifies the building as a shelter.

1.8.4.2 Signage at Schools and Places of Work. Signage for shelters at schools and places of work should be clearly posted and should direct occupants through the building or from building to building. If the shelter is in a government-funded or public-funded facility, a placard should be placed on the outside of the building designating it an emergency shelter (see Figure 1-12). It is recommended that signage be posted on the outside of all other types of shelters as well. It is important to note, however, that once a public building has been identified as a shelter, people who live or work in the area around the shelter will expect the shelter to be open during an event. Shelter owners should be aware of this and make it clear that the times when a shelter will be open may be limited. For example, a shelter in an elementary school or commercial building may not be accessible at night.

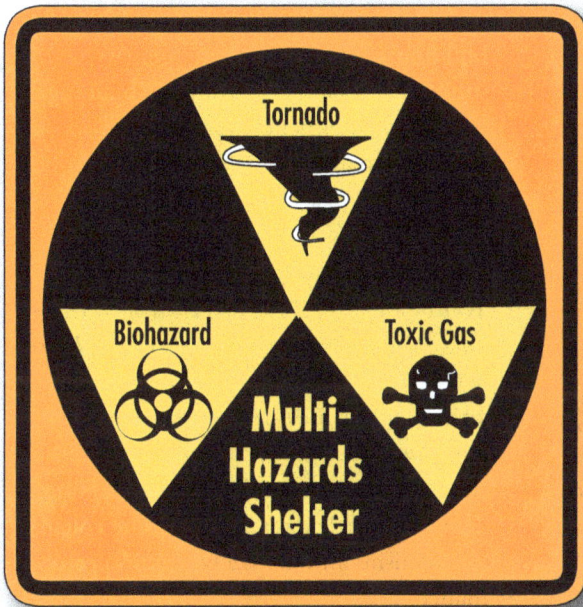

Figure 1-12   Shelter signage

1.9 EVACUATION CONSIDERATIONS

When designing a shelter, evacuation is one of the most important aspects to save lives. During the attack of the World Trade Center, good and well-marked egress was critical for thousands of people to evacuate the buildings. The same concept is applicable to shelters. Good ingress and egress, along with robustness and redundancy of the structural system, is critical for a sound design.

The matter of high-rise evacuation has become vital since September 11, 2001, as a result of the fatalities of almost 3,000 building occupants and emergency personnel. Life safety is provided to building occupants by either giving them the opportunity to evacuate to a safer place or be protected in place.

> Every building should have an emergency evacuation and shelter-in-place plan that is coordinated with the local community emergency manager. Building stakeholders and tenants should develop the plan with the objective to save lives and property, and to recover and restore the business should an event occur. The NFPA 1600 *Standard on Disaster/Emergency Management and Business Continuity Programs* publication provides a framework and recommendations for developing a plan. The building owner, property manager, and tenants should work with the local community to develop an evacuation versus shelter-in-place options matrix as shown in Table 1-6.

The National Institute of Standards and Technology (NIST) *Final Report of the National Construction Safety Team on the Collapses of the World Trade Center Towers* conducted analysis of the life safety systems and emergency response to validate and expand the state of the practice for high-rise buildings. The NIST study was focused on the collapse mechanisms and life safety systems performance.

As a result of the collapse of the World Trade Center towers, NIST identified three major scenarios that are not considered adequately in current design practice:

○ Frequent but low severity events (for design of sprinkler system)

○ Moderate but less frequent events (for design of compartmentation)

○ A maximum credible fire (for design of passive fire protection on the structure)

Table 1-6: Evacuation Versus Shelter-in-place Options Matrix

Attack Agent	Timeframe and Protection Objective	Occupant/Personnel Action
Chemical – Exterior Release	Immediate - shelter in safe room, minimize duration and concentration exposure	Use portable air filtration, wait for first responder extraction
Chemical – Interior Release	Immediate - don PPE and evacuate, minimize duration and concentration exposure	Move perpendicular to plume direction, seek decontamination and medical treatment
Biological – Exterior Release	Immediate - shelter in safe room, do not touch agents, use time to advantage to identify safe evacuation route	Use portable air filtration, wait for first responder extraction, seek decontamination and medical treatment
Biological – Interior Release	Immediate - don PPE and evacuate, minimize duration and concentration exposure	Seek decontamination and medical treatment
Radiological/Nuclear – Exterior Release	Immediate - shelter in safe room, minimize duration and concentration exposure	Use portable air filtration, wait for first responder extraction
Radiological/Nuclear – Interior Release	Immediate - don PPE and evacuate, minimize duration and concentration exposure	Seek decontamination and medical treatment
Explosive Blast - Exterior	Immediate - shelter in safe room	Use portable air filtration, wait for first responder extraction
Explosive Blast - Interior	Immediate - don PPE and evacuate	Seek medical treatment

Three methods are followed for the evacuation of buildings. One method consists of evacuating all occupants simultaneously. Alternatively, occupants may be evacuated in phases, where the floor levels closest to the event are evacuated first and then other floor levels are evacuated on an as needed basis. Phased evacuation is instituted to permit people on the floor levels closest to the threatening hazard to enter the stairway unobstructed by queues formed by people from all other floors also being in the stairway. Those who are below the emergency usually are encouraged to stay in place until the endangered people from above are already below this floor level.

The concept of occupant relocation to other floors is usually the best course of action for many types of building emergencies. This method normally involves movement of occupants, from the fire floor, the floor above, and floor below to a lower level until the danger passes.

Evacuation involves providing people with the means to exit the building. The egress system involves the following considerations:

Capacity. A sufficient number of exits of adequate width to accommodate the building population need to be provided to allow occupants to evacuate safely.

Access. Occupants need to be able to access an exit from wherever the fire is, and in sufficient time prior to the onset of untenable conditions. Alternative exits should be remotely located so that all exits are not simultaneously blocked by a single incident.

Exit Design. Exits need to be separated from all other portions of the building in order to provide a protected way of travel to the exit discharge. This involves designing to preclude fire and smoke from entering the exit and will usually involve structural stability.

In general, the means of egress system is designed so that occupants travel from the office space along access paths such as corridors or aisles until they reach the exit or a safer place. An exit is commonly defined as a protected path of travel to the exit discharge (NFPA 101, 2006). The stairways in a high-rise building commonly meet the definition of an exit. In general, the exit is intended to provide a continuous, unobstructed path to the exterior or to another area that is considered safe. Most codes require that exits discharge directly to the outside. Some codes, such as NFPA 101, permit up to half of the exits to discharge within the building, given that certain provisions are met.

There is no universally accepted standard on emergency evacuation. Design considerations for high-rise buildings relative to these two options involve several aspects, including design of means of egress, the structure, and active fire protection systems (e.g., detection and alarm, suppression, and smoke management). Many local jurisdictions, through their fire department public education programs, have developed comprehensive and successful evacuation planning models but, unless they are locally adopted, there is no legal mandate to exercise the plans. Seattle, Phoenix, Houston, and Portland, Oregon, are among the cities that have developed comprehensive programs.

The NIST life safety, egress, and emergency response findings provide valuable lessons learned for future shelter evacuation design. Currently, building fire protection is based on a four-level hierarchical strategy comprising alarm and detection, suppression (sprinklers and firefighting), compartmentation, and passive protection of the structure.

○ Manual stations and detectors are typically used to activate fire alarms and notify building occupants and emergency services.

○ Sprinklers are designed to control small and medium fires and to prevent fire spread beyond the typical water supply design area of about 1,500 square feet.

○ Compartmentation mitigates the horizontal spread of more severe but less frequent fires and typically requires fire-rated partitions for areas of about 12,000 square feet. Active firefighting measures also cover up to about 5,000 square feet to 7,500 square feet.

○ Passive protection of the structure seeks to ensure that a maximum credible fire scenario, with sprinklers compromised or overwhelmed and no active firefighting, results in burnout, not overall building collapse. The intent of building codes is also for the building to withstand local structural collapse until occupants can escape and the fire service can complete search and rescue operations.

NIST recommends that building evacuation should be improved to include system designs that facilitate safe and rapid egress, methods for ensuring clear and timely emergency communications to occupants, better occupant preparedness for evacuation during emergencies, and incorporation of appropriate egress technologies. When designing good evacuation systems and routes of ingress and egress, designers should take into account the following considerations:

○ As stated above, improved building evacuation, including system designs that facilitate safe and rapid egress, methods for ensuring clear and timely emergency communications to occupants, better occupant preparedness for evacuation during emergencies, and incorporation of appropriate egress technologies. Primary and secondary evacuation routes and exits should be designated and clearly marked and well lit. Signs should be posted.

○ Improved emergency response, including better access to the buildings and better operations, emergency communications, and command and control in large-scale emergencies.

○ Emergency lighting should be installed in case a power outage occurs during an evacuation.

Recommendation 16. NIST recommends that public agencies, non-profit organizations concerned with building and fire safety, and building owners and managers should develop and carry out public education campaigns, jointly and on a nationwide scale, to improve building occupants' preparedness for evacuation in case of building emergencies.

Recommendation 17. NIST recommends that tall buildings should be designed to accommodate timely full building evacuation of occupants due to building-specific or large-scale emergencies such as widespread power outages, major earthquakes, tornadoes, hurricane without sufficient advanced warning, fires, accidental explosions, and terrorist attacks. Building size, population, function, and iconic status should be taken into account in designing the egress system. Stairwell and exit capacity should be adequate to accommodate counterflow due to emergency access by responders.

Recommendation 18. NIST recommends that egress systems should be designed: (1) to maximize remoteness of egress components (i.e., stairs, elevators, exits) without negatively impacting the average travel distance; (2) to maintain their functional integrity and survivability under foreseeable building-specific or large-scale emergencies; and (3) with consistent layouts, standard signage, and guidance so that systems become intuitive and obvious to building occupants during evacuations.

Recommendation 19. NIST recommends that building owners, managers, and emergency responders develop a joint plan and take steps to ensure that accurate emergency information is communicated in a timely manner to enhance the situational awareness of building occupants and emergency responders affected by an event. This should be accomplished through better coordination of information among different emergency responder groups, efficient sharing of that information among building occupants and emergency responders, more robust design of emergency public address systems, improved emergency responder communication systems, and use of the Emergency Broadcast System (now known as the Integrated Public Alert and Warning System) and Community Emergency Alert Networks.

Recommendation 21. NIST recommends the installation of fire-protected and structurally hardened elevators to improve emergency response activities in tall buildings by providing timely emergency access to responders and allowing evacuation of mobility-impaired building occupants. Such elevators should be installed for exclusive use by emergency responders during emergencies. In tall buildings, consideration also should be given to installing such elevators for use by all occupants.

○ Evacuation routes and emergency exits should be:

    ○ wide enough to accommodate the number of evacuating personnel.

    ○ clear and unobstructed at all times.

    ○ unlikely to expose evacuating personnel to additional hazards.

○ Evacuation routes should be evaluated by a professional.

The 2006 editions of NFPA 101, *Life Safety Code* and NFPA 5000, *Building Construction and Safety Code* addressed the issue of counterflow between first responders and descending occupants. The new provisions mandate a minimum stair width of 56 inches (1,420 mm) when a stair is designed to handle an aggregate or accumulated of 2,000 or more occupants. Previous criteria required 44 inches (1,120 mm) minimum.

It is also important to designate assembly areas and a means of obtaining an accurate account of personnel after a site evacuation. Designate areas where personnel should gather after evacuating (see Section 1.10). A head count should be taken after the evacuation. The names and last known locations of personnel not accounted for should be determined and given to the Emergency Operations Coordinator (EOC). (Confusion in the assembly areas can lead to unnecessary and dangerous search and rescue operations.) A method for accounting for non-employees (e.g., suppliers and customers) should also be established.

In addition, procedures should be established for further evacuation in case the incident expands. This may consist of sending employees home by normal means or providing them with transportation to an off-site location.

1.10 KEY OPERATIONS ZONES

Key operations zones refer to the shelter site surrounding areas and entry and exit control points that need to be taken into consideration when designing a shelter.

For catastrophic incidents depicted in the planning scenarios related to the National Response Plan - Catastrophic Incident Supplement (NRP-CIS), decontamination involves several

DESIGN CONSIDERATIONS

related and sequential activities. Chief among these are (1) immediate (or gross) decontamination of persons exposed to toxic/hazardous substances; (2) continual decontamination of first responders so that they can perform their essential functions; (3) decontamination of animals in service to first responders; (4) continual decontamination of response equipment and vehicles; (5) secondary, or definitive, decontamination of victims at medical treatment facilities to enable medical treatment and protect the facility environment; (6) decontamination of facilities (public infrastructure, business and residential structures); and (7) environmental (outdoor) decontamination supporting recovery and remediation.

1.10.1 Containment Zones

There are three zones of containment after an event:

○ Hot Zone (the area where the agent or contamination is in high concentration and high exposure, typically an ellipse or cone extending downwind from the release)

○ Warm Zone (the area where the agent or contamination is in low concentration or minimal exposure, typically a half circle in the above wind direction)

○ Cold Zone (those areas outside of the Hot and Warm zones that have not been exposed to the agent or contamination)

The three zones and staging areas are shown in Figure 1-13.

Figure 1-13
Operations Zones, Casualty
Collection Point (CCP), and
Safe Refuge Area (SRA)

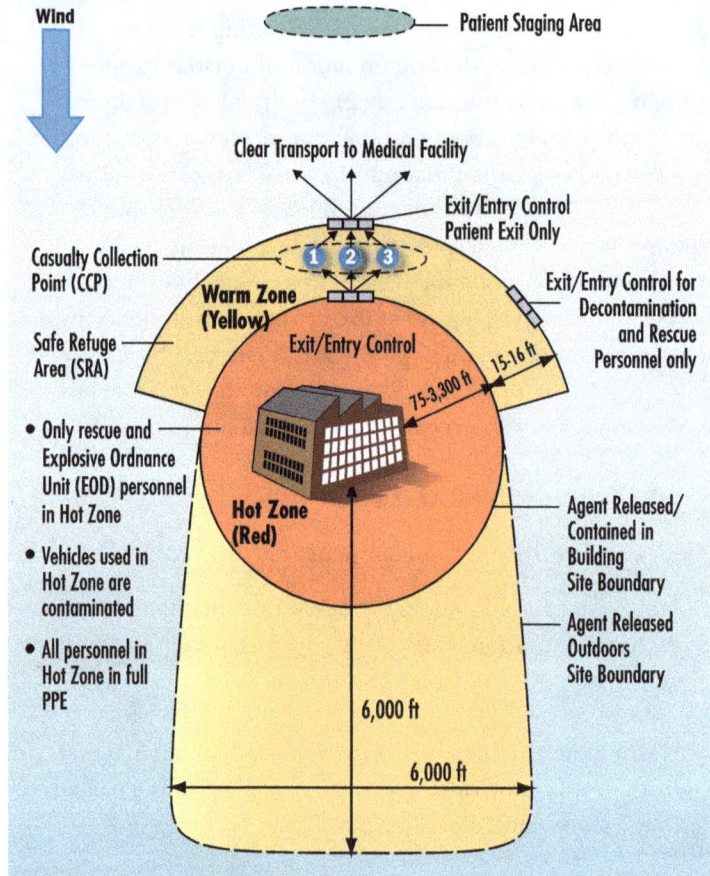

Wind

Patient Staging Area

Clear Transport to Medical Facility

Exit/Entry Control
Patient Exit Only

Casualty Collection
Point (CCP)

Warm Zone
(Yellow)

Exit/Entry Control for
Decontamination
and Rescue
Personnel only

Safe Refuge
Area (SRA)

Exit/Entry Control

15-16 ft

75-3,300 ft

• Only rescue and
Explosive Ordnance
Unit (EOD) personnel
in Hot Zone

Hot Zone
(Red)

Agent Released/
Contained in
Building
Site Boundary

• Vehicles used in
Hot Zone are
contaminated

Agent Released
Outdoors
Site Boundary

• All personnel in
Hot Zone in full
PPE

6,000 ft

6,000 ft

Casualty Collection Point (CCP)

① **Litter Decontamination**
Non-ambulatory Delayed
Treatment

② **Litter Decontamination**
Immediate Treatment

③ **Ambulatory
Decontamination** Minimal
Treatment Ambulatory
Delayed Treatment

• Mass decontamination occurs in
the Warm Zone

• Safe refuge area in the Warm
Zone used to assemble
individuals who are witnesses to
the incident and separation of
contaminated from
non-contaminated persons

Shelter occupants should not leave the shelter until rescue personnel arrive to escort occupants to the Cold Zone. The building occupants must go through several staging areas to ensure that any CBR contamination is not spread across a larger geographical area. There are two processes currently used to evacuate an area; the Ladder Pipe Decontamination System (LDS, see Figure 1-14) and the Emergency Decontamination Corridor System (EDCS, see Figure 1-15).

Ladder Pipe Decontamination System (LDS)

Advantages

- Rapid setup time
- Provides large capacity high volume low pressure shower
- Rapid hands-free mass decontamination

Disadvantages

- No privacy
- Increased chance of hypothermia from exposure to elements

Composed of:

- Ladder pipe/truck
- 2 engines
- Hand-held hose lines

Setup:

- Engines placed approximately 20 feet apart
- 2 1/2-inch fog nozzles set at wide fog pattern attached to pump discharges
- Truck with fog nozzle placed on ladder pipe to provide downward fog pattern

Firefighters (FFs) can be positioned at either or both ends of the shower area to apply additional decontamination wash

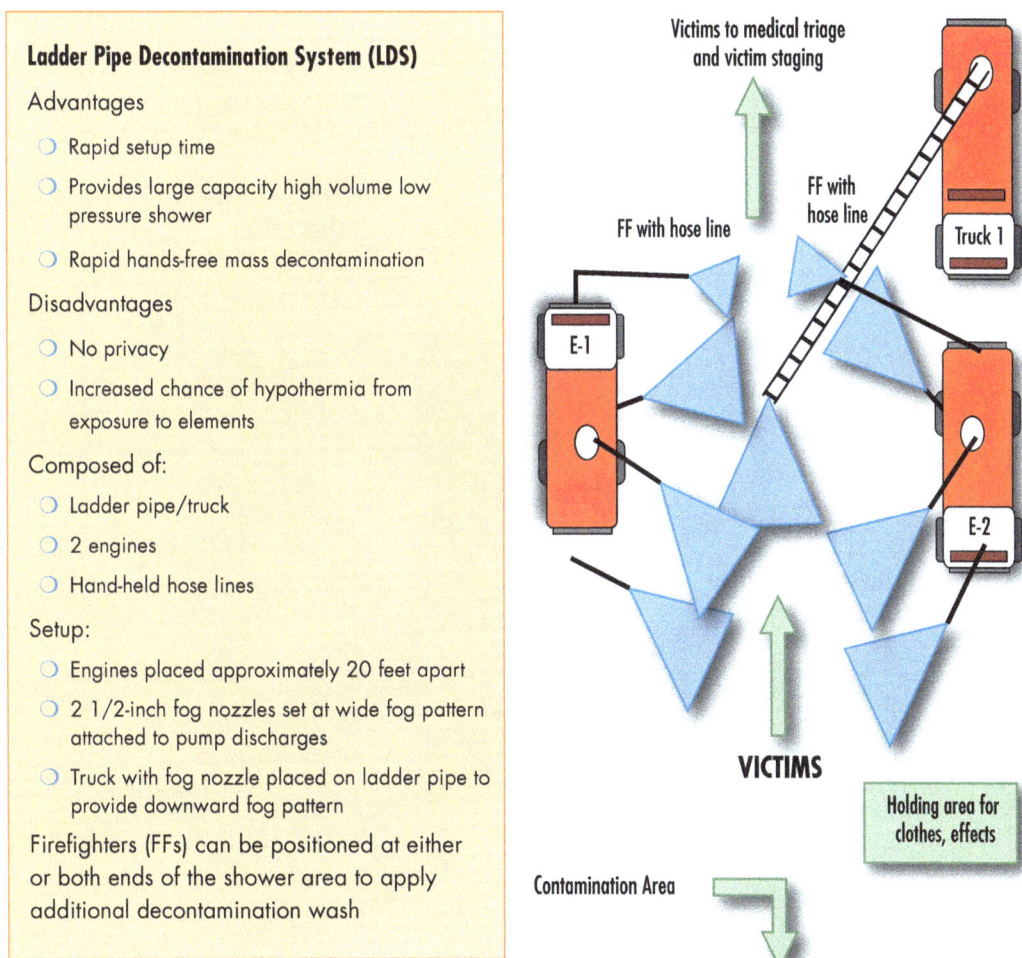

Figure 1-14   NRP-CIS Ladder Pipe Decontamination System (LDS)

SOURCE: NRP-CIS

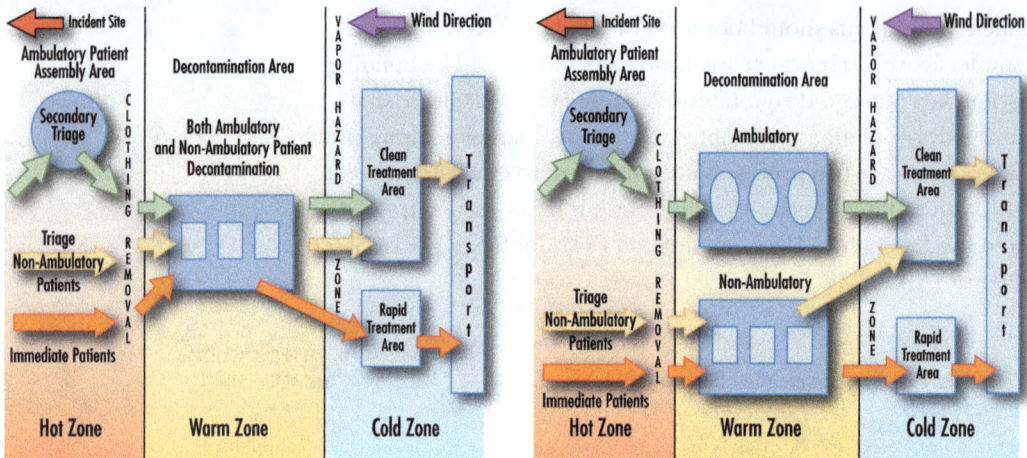

EDCS Layouts

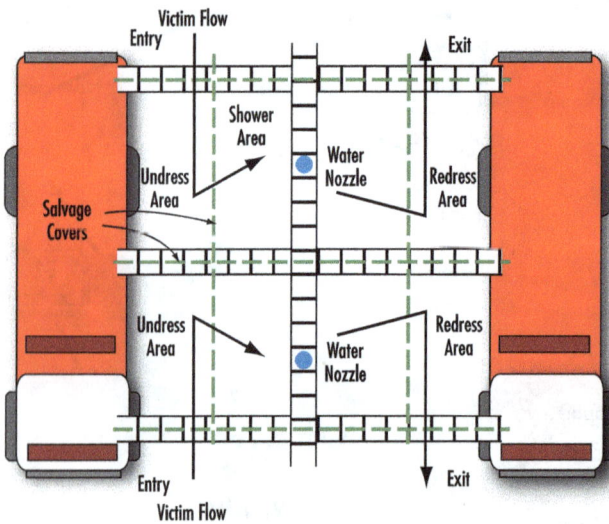

EDCS Decomtamination Area Setup

Figure 1-15
NRP-CIS Emergency Decontamination Corridor System (EDCS)

SOURCE: NRP-CIS

1.10.2 Staging Areas and Designated Entry and Exit Access Control Points

To control the potential spread of a CBRE agent and ensure the safety of the victims and first responders, the Incident Commander (IC) will establish several staging areas and designated entry and exit access control points for the three zones.

○ **Patient Staging Area (PSA).** The PSA is located in the Cold Zone and is the transfer point for victims that have been stabilized for transport to higher care medical facilities or for fatalities to be transported to morgue facilities (see Figures 1-16 and 1-17). The PSA area must be large enough to accommodate helicopter operations and a large number of ambulances.

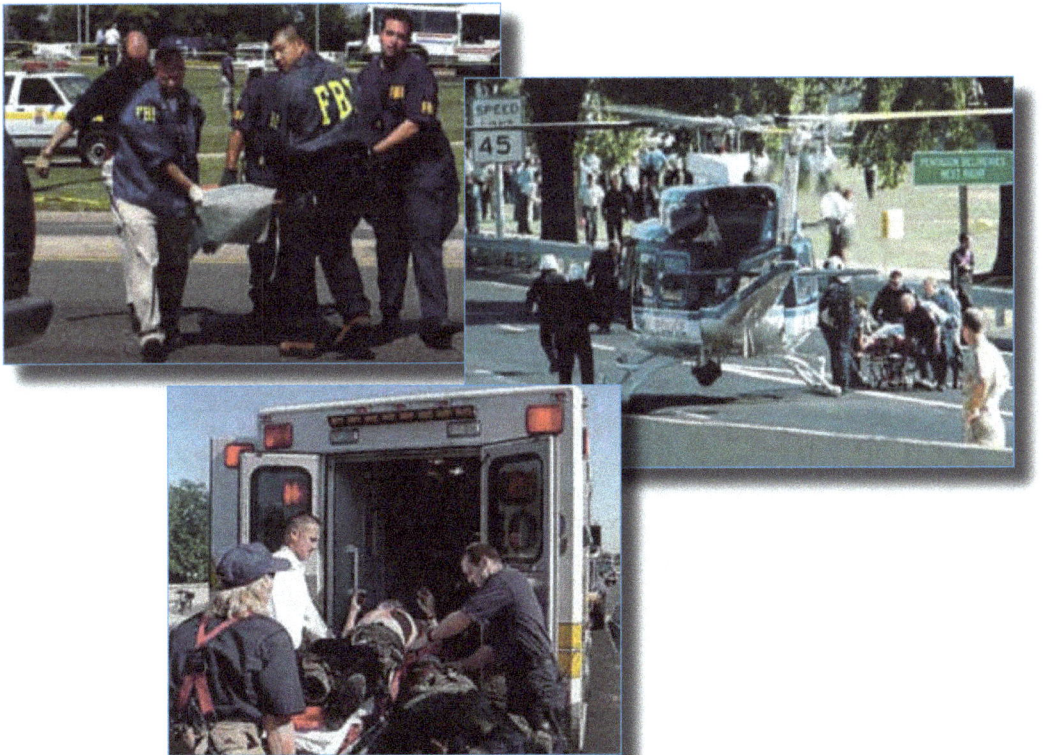

Figure 1-16   Patient staging area and remains recovery

SOURCE: ARLINGTON COUNTY AFTER-ACTION REPORT

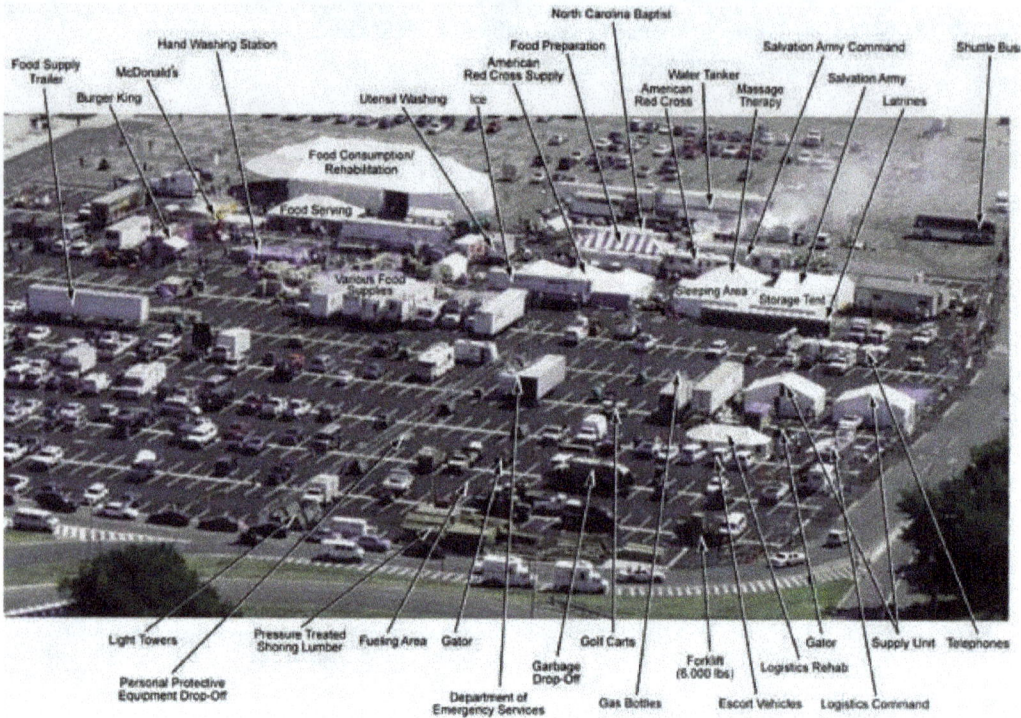

Figure 1-17   Example of Pentagon staging and recovery operations

SOURCE: ARLINGTON COUNTY AFTER-ACTION REPORT

○ **Contamination Control Area (CCA).** The CCA (see Figure 1-18) is located on the boundary of the Cold Zone and Warm Zone and used by the rescue and decontamination personnel to enter and exit the Warm Zone. There are several processing stations, a resupply and refurbishment area, and a contaminated waste storage area. Mass casualty decontamination occurs in the Warm Zone.

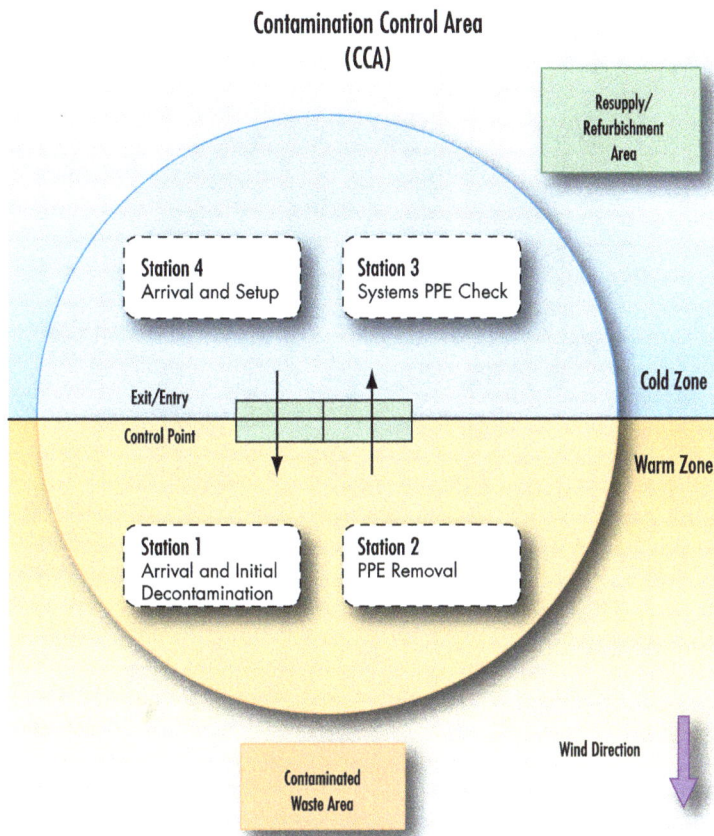

Contamination Control Area (CCA)

Resupply/Refurbishment Area

Station 4
Arrival and Setup

Station 3
Systems PPE Check

Cold Zone

Exit/Entry
Control Point

Warm Zone

Station 1
Arrival and Initial Decontamination

Station 2
PPE Removal

Wind Direction

Contaminated Waste Area

Figure 1-18
Contamination Control Area (CCA)

○ **Safe Refuge Area (SRA).** The SRA is located in the Warm Zone and used to assemble survivors and witnesses that are not injured and will require minimal medical attention and decontamination. Law enforcement and FBI agents can conduct interviews and gather evidence at the SRA.

Evidence collection can occur in any of the three zones as shown in Figure 1-19.

Figure 1-19
Site and evidence
collection on the site

SOURCE: ARLINGTON COUNTY
AFTER-ACTION REPORT

Designated entry/exit access control points will be between each
of the zones. The entry/exit access control point between the Hot
and Warm Zones is used by the first responders in PPE to enter/
exit the Hot Zone and extract victims and casualties (both contaminated and uncontaminated) to the Warm Zone. The patient
entry/exit access control point between the Warm and Cold Zones
is used as a one-way exit out to move decontaminated uninjured
personnel and medically stabilized casualties. The first responders
entry/exit between the Cold and Warm Zones is used by the first
responders, rapid visualization and structural evaluation team,
and debris operations personnel to enter and exit the site, and includes equipment shown in Figure 1-20.

Figure 1-20   Rescue team coordination prior to entering a site
SOURCE: ARLINGTON COUNTY AFTER-ACTION REPORT

Between the Hot Zone entry access control point and the patient exit control point, there will be a casualty collection point. The CCP (Figure 1-13) is located in the Warm Zone and will typically have three processing stations:

○ Station 1 – Litter decontamination and non-ambulatory delayed treatment patients

○ Station 2 – Litter decontamination and immediate treatment patients

○ Station 3 – Ambulatory decontamination, minimal treatment patients, and ambulatory delayed treatment patients

STRUCTURAL DESIGN CRITERIA   2

2.1 OVERVIEW

This chapter discusses explosive threat parameters and measures needed to protect shelters from blast effects. Structural systems and building envelope elements for new and existing shelters are analyzed; shelters and FEMA model building types are discussed; and protective design measures for the defined building types are provided, as are design guidance and retrofit issues. The purpose of this chapter is to offer comprehensive information on how to improve the resistance of shelters when exposed to blast events.

2.2 EXPLOSIVE THREAT PARAMETERS

A detonation involves supersonic combustion of an explosive material and the formation of a shock wave. The three parameters that primarily determine the characteristics and intensity of blast loading are the weight of explosives, the type of the explosives, and the distance from the point of detonation to the protected building. These three parameters will primarily determine the characteristics and intensity of the blast loading. The distance of the protected building from the point of explosive detonation is commonly referred to as the stand-off distance. The critical locations for detonation are taken to be at the closest point that a vehicle can approach, assuming that all security measures are in place. Typically, this would be a vehicle parked along the curb directly outside the facility, or at the vehicle access control gate where inspection takes place. Similarly, a critical location may be the closest point that a hand carried device can be deposited.

There is also no way to effectively know the size of the explosive threat. Different types of explosive materials are classified as High Energy and Low Energy and these different classifications greatly influence the damage potential of the detonation. High Energy explosives, which efficiently convert the material's chemical energy into blast pressure, represent less than 1 percent of all explosive detonations reported by the FBI Bomb Data Center. The vast majority of incidents involve Low Energy devices in which a significant portion of the explosive material is consumed by deflagration, which is a process of subsonic combustion that usually propagates through thermal conductivity and is typically less destructive than a detonation. In these cases, a large portion of the material's chemical energy is dissipated as thermal energy, which may cause fires or thermal radiation damage.

For a specific type and weight of explosive material, the intensity of blast loading will depend on the distance and orientation of the blast waves relative to the protected space. A shock wave is characterized by a nearly instantaneous rise in pressure that decays exponentially within a matter of milliseconds, which is followed by a longer term but lower intensity negative phase. The initial magnitude of pressure is termed the peak pressure and the area under a graph of pressure plotted as a function of time, also known as the airblast pressure time history, is termed the impulse (see Figure 2-1). Therefore, the impulse associated with the shock wave considers both the pressure intensity and the pulse duration. As the front of the shock-wave propagates away from the source of the detonation at supersonic speed, it expands into increasingly larger volumes of air; the peak incident pressure at the shock front decreases and the duration of the pressure pulse increases. The magnitude of the peak pressures and impulses are reduced with distance from the source and the resulting patterns of blast loads appear to be concentric rings of diminishing intensity. This effect is analogous to the circular ripples that are created when an object is dropped in a pool of water. The shock front first impinges on the leading surfaces of a building located within its path and is reflected and diffracted, creating focus and shadow zones on the building envelope. These patterns of blast load intensity are complicated as the waves engulf the entire building. The pressures that load the roof, sides, and rear of the building are termed incident pressures, while the pressures that load the building envelope directly opposite the explosion are termed reflected pressures. Both the intensity of peak pressure and the impulse

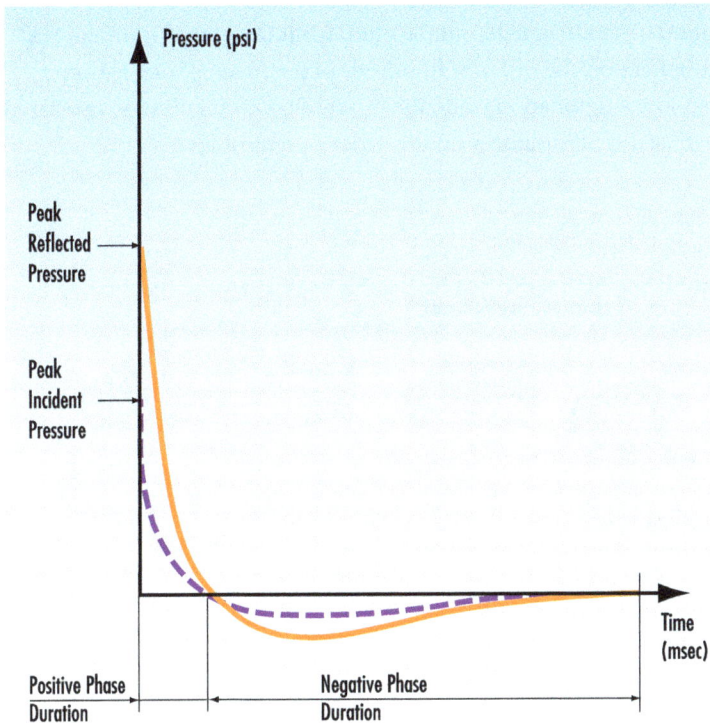

Figure 2-1
Airblast pressure time
history

affect the hazard potential of the blast loading. A detailed analysis is required to determine the magnitude of pressure and impulse that may load each surface relative to the origin of the detonation.

The thresholds of different types of injuries associated with damage to wall fragments and/or glazing are depicted in Figure 2-2. This range to effects chart shows a generic interaction between the weight of the explosive threat and its distance to an occupied building. These generic charts, for conventional construction, provide information to law enforcement and public safety officials that allow them to establish safe evacuation distances should an explosive device be suspected or detected. However, these distances are so site-specific that the generic charts provide little more than general guidance in the absence of more reliable site-specific information. Based on the information provided in the chart, the onset of significant glass debris hazards is associated with stand-off distances on the order of hundreds of feet from a vehicle-borne explosive detonation while the onset of column failure is associated with stand-off distances on the order of tens of feet.

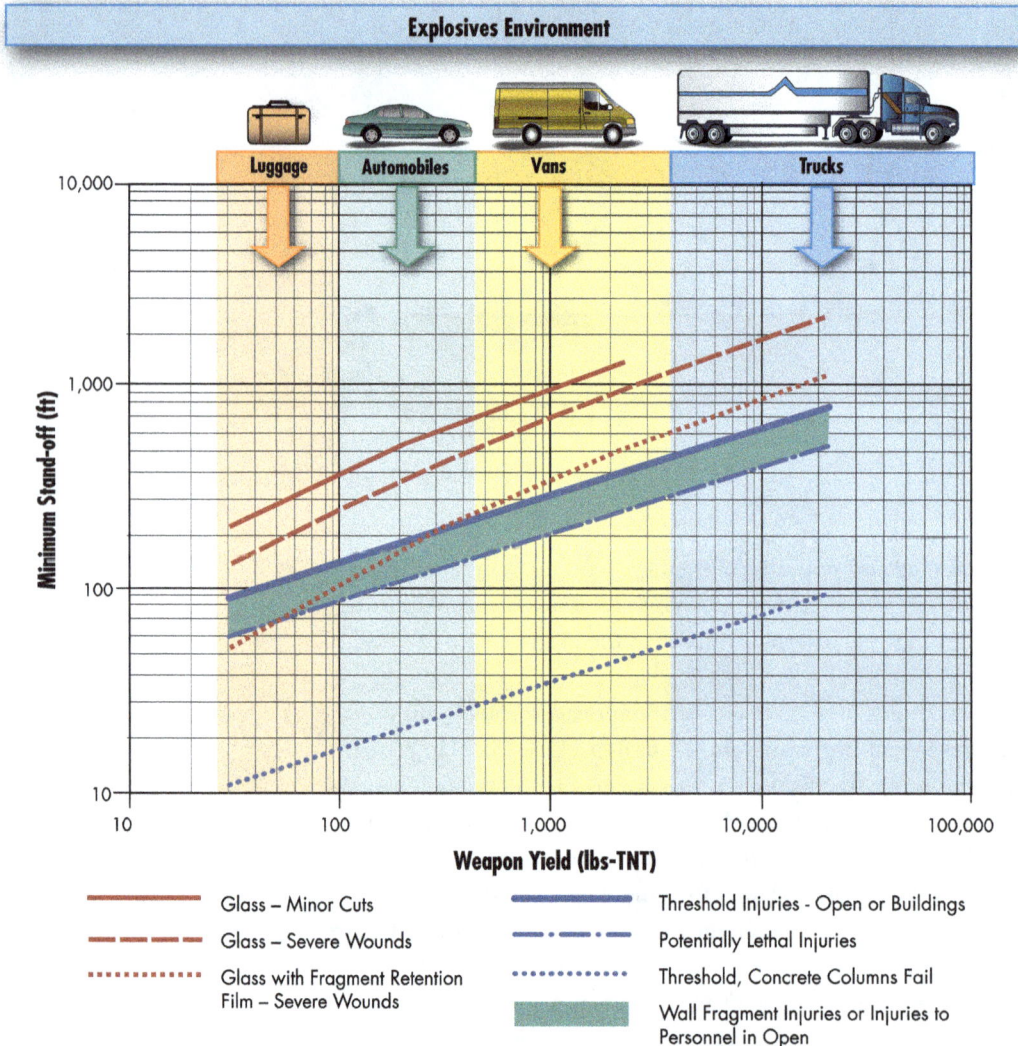

Figure 2-2   Range to effects chart

SOURCE: DEFENSE THREAT REDUCTION AGENCY

STRUCTURAL DESIGN CRITERIA

2.2.1 Blast Effects in Low-rise Buildings

Many shelters can be part of low-rise buildings. Although small weights of explosives are not likely to produce significant blast loads on the roof, low-rise buildings may be vulnerable to blast loadings resulting from large weights of explosives at large stand-off distances that may sweep over the top of the building. The blast pressures that may be applied to these roofs are likely to far exceed the conventional design loads and, unless the roof is a concrete deck or concrete slab structure, it may fail. There is little that can be done to increase the roof's resistance to blast loading that doesn't require extensive renovation of the building structure. Figure 2-3 shows the ever expanding blast wave as it radiates from the point of detonation and causes, in sequence of events, the building envelope to fail, the internal uplift on the floor slabs, and eventually the engulfment of the entire building.

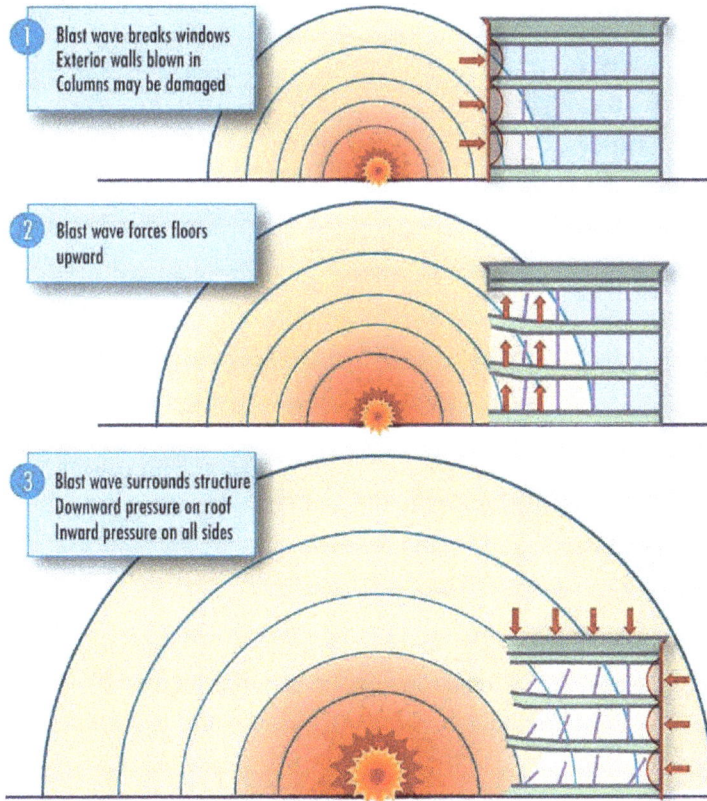

Figure 2-3

Blast damage

SOURCE: NAVAL FACILITIES ENGINEERING SERVICE CENTER, *USER'S GUIDE ON PROTECTION AGAINST TERRORIST VEHICLE BOMBS*, MAY 1998

In addition to the blast pressures that may be directly applied to the exterior columns and spandrel beams, the forces collected by the building envelope will be transferred through the slabs to the structural frame or shear walls that transfer lateral loads to the foundations. The extent of damage will be greatest in close proximity to the detonation; however, depending on the intensity of the blast, large inelastic deformations will extend throughout the building and cause widespread cracking to structural and nonstructural partitions.

In addition to the hazard of impact by building envelope debris propelled into the building or roof damage that may rain down, the occupants may also be vulnerable to much heavier debris resulting from structural damage. Progressive collapse occurs when an initiating localized failure causes adjoining members to be overloaded and fail, resulting in a cascading sequence of damage that is disproportionate to the originating extent of localized failure. The initiating localized failure may result from a sufficiently sized parcel bomb that is in contact with a critical structural element or from a vehicle sized bomb that is located a short distance from the building (see Figure 2-4). However, a large explosive device at a large stand-off distance is not likely to selectively fail a single structural member; any damage that results from this scenario is more likely to be widespread and the ensuing collapse cannot be considered progressive. Although progressive collapse is not typically an issue for buildings three stories or shorter, transfer girders and non-ductile, non-redundant construction may produce structural systems that are not tolerant of localized damage conditions. The columns that support transfer girders and the transfer girders themselves may be critical to the stability of a large area of floor space.

Figure 2-4
Alfred P. Murrah Federal
Office Building
SOURCE: U.S. AIR FORCE

STRUCTURAL DESIGN CRITERIA

As an example, panelized construction that is sufficiently tied together can resist the localized damage or large structural deformations that may result from an explosive detonation. Although the explosive detonation opposite the Khobar Towers destroyed the exterior façade, the panelized structure was sufficiently tied together to permit relatively large deformations without loss of structural stability (see Figure 2-5). This highlights the benefits of ductile and redundant detailing for all types of construction.

Figure 2-5
Khobar Towers
SOURCE: U.S. AIR FORCE

To mitigate the effects of in-structure shock that may result from the infilling of blast pressures through damaged enclosures, nonstructural overhead items should be located below the raised floors or tied to the ceiling slabs with seismic restraints. Nonstructural building components, such as piping, ducts, lighting units, and conduits must be sufficiently tied back to the building to prevent failure of the services and the hazard of falling debris.

The contents of this manual supplement the information provided in FEMA 361, *Design and Construction Guidance for Community Shelters* and FEMA 320, *Taking Shelter From the Storm: Building a Safe Room Inside Your House.* Although this publication does not specifically address nuclear explosions and shelters that protect against radiological fallout, this information may be found in FEMA TR-87, *Standards for Fallout Shelters.* The contents of FEMA 452, *A How-To Guide to Mitigate Potential Terrorist Attacks Against Buildings* will help the reader identify critical assets and functions within buildings, determine the threats to these assets, and assess the vulnerabilities associated with those threats.

2.2.2   Blast Effects in High-rise Buildings: The Urban Situation

High-rise buildings must resist significant gravity and lateral load effects; although the choice of framing system and specific structural details will determine the overall performance, the lower floors, which are in closest proximity to a vehicle-borne threat, are inherently robust and more likely to be resistant to blast loading than smaller buildings. However, tall buildings are likely to be located in dense urban environments that tend to trap the blast energy within the canyon like streets as the blast waves reflect off of neighboring structures. Furthermore, tall buildings are likely to contain underground parking and loading docks that can introduce significant internal explosive threats. While these internal threats may be prevented through rigorous access control procedures, there are few conditions in which vehicular traffic can be restricted on city streets. Anti-ram streetscape elements are required to maintain a guaranteed stand-off distance from the face of the building.

In addition to the hazard of structural collapse, the façade is a much more fragile component. While the lower floor façade is likely to fail in response to a sizable vehicle threat at a sidewalk's distance from the building, the peak pressures and impulses at higher elevations diminish due to the increased stand-off distance and the associated shallow angle of incidence (measured with respect to the vertical height of the building). Although reflections off of neighboring structures are likely to affect the intensity of blast loads, the façade loads at the upper floors will be considerably lower than the loads at the lower floors and the extent of façade debris will reflect this. A detailed building-specific analysis of the structure and the façade is required to identify the inherent strengths and vulnerabilities. This study will indicate the safest place to locate the shelter.

2.3 HARDENED CONSTRUCTION

2.3.1   Structural System

A shelter will only be effective if the building in which it is located remains standing. It is unreasonable to design a shelter within a building with the expectation that the surrounding structure may collapse. Although the shelter must be able to resist debris impact, it is not reasonable for it to withstand the weight of the building crashing down upon it. Therefore, the effectiveness of the shelter will depend on the ability of the building to sustain damage, but remain standing. The ability of a building to withstand an explosive event and remain standing depends on the characteristics of the structure. Some of these characteristics include:

○ **Mass**. Lightweight construction may be unsuitable for providing resistance to blast loading. Inertial resistance may be required in addition to the strength and ductility of the system.

○ **Shear capacity.** Shear is a brittle mode of failure and primary members and/or their connections should therefore be designed to prevent shear failure prior to the development of the flexural capacity.

○ **Capacity for resisting load reversals.** In response to sizable blast loads, structural elements may undergo multiple cycles of large deformation. Similarly, some structural elements may be subjected to uplift pressures, which oppose conventional gravity load design. The effects of rebound and uplift therefore require blast-resistant members to be designed for significant load reversals. Depending on the cable profile, pre-tensioned or post-tensioned construction may provide limited capacity for abnormal loading patterns and load reversals. Draped tendon systems designed for gravity loads may be problematic; however, the higher quality fabrication and material properties typical for precast construction may provide enhanced performance of precast elements designed and detailed to resist uplift and rebound effects resulting from blast loading. Seated connection systems for steel and precast concrete systems must also be designed and detailed to accommodate uplift forces and rebound resulting from blast loads. The use of headed studs is recommended for affixing concrete fill over steel deck to beams for uplift resistance.

○ **Redundancy.** Multiple alternative load paths in the vertical-load-carrying system allow gravity loads to redistribute in the event of failure of structural elements.

○ **Ties.** An integrated system of ties in perpendicular directions along the principal lines of structural framing can serve to redistribute loads during catastrophic events.

○ **Ductility.** Structural members and their connections may have to maintain their strength while undergoing large deformations in response to blast loading. The ability of a member to develop inelastic deformations allows it to dissipate considerable amounts of blast energy. The ratio of a member's maximum inelastic deformation to a member's elastic limit is a measure of its ductility. Special detailing is required to enable buildings to develop large inelastic deformations (see Figure 2-6).

Historically, cast-in-place reinforced concrete was the preferred material for explosion-mitigating construction. This is the material used for military bunkers, and the military has performed extensive research and testing of its performance. Among its benefits, reinforced concrete has significant mass, which improves its inertial resistance; it can be readily proportioned for ductile behavior and may be detailed to achieve continuity between members. Finally, concrete columns are less susceptible to global buckling in the event of the loss of a floor system. However, steel may be similarly detailed to take advantage of its inherent ductility and connections may be designed to provide continuity between members. Similarly, panelized precast concrete systems can be detailed to permit significant deformations in response to explosive loading, as demonstrated by the performance of Khobar Towers.

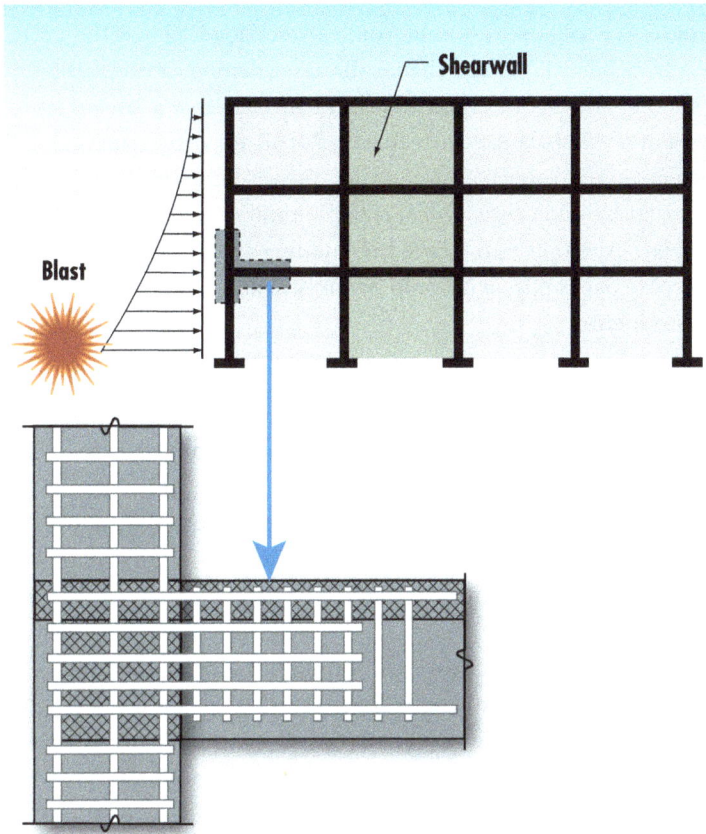

Figure 2-6
Ductile detailing of
reinforced concrete
structures

Protective design further requires the system to accept localized failure without precipitating a collapse of a greater extent of the structure. By allowing the building to bridge over failed components, building robustness is greatly improved and the unintended consequences of extreme events may be mitigated. However, it may not be possible for existing construction to be retrofitted to limit the extent of collapse to one floor on either side of a failed column. If the members are retrofitted to develop catenary behavior, the adjoining bays must be upgraded to resist the large lateral forces associated with this mode of response. This may require more extensive retrofit than is either feasible or desirable. In such a situation, it may be desirable to isolate the collapsed region rather than risk propagating the collapse to adjoining bays. The retrofit of existing buildings to protect against a potential progressive collapse resulting from extreme loading may therefore best be achieved through the localized hardening of vulnerable columns. These columns need only be upgraded to a level of resistance that balances the capacities of all adjacent structural elements. At greater blast intensities, the resulting damage would be extensive and create global collapse rather than progressive collapse. Attempts to upgrade the building to conform to the alternate path approach would be invasive and potentially counterproductive.

2.3.2 Loads and Connections

Because the shelter will likely suffer significant damage in response to extreme loading conditions, the shelter must be able to withstand both the direct loading associated with the natural or manmade hazard and the debris associated with the damaged building within which it is housed.

Structural systems that provide a continuous load path that supports all vertical and lateral loads acting on a building are preferred. A continuous load path ties all structural components together and the fasteners used in the connections must be capable of developing the full capacity of the members. In order to provide comprehensive protection, the capacity of each component must be balanced with the capacity of all other components and the connection details that tie them together. Because all applied loads must eventually be transferred to the foundations, the load path must be continuous from the uppermost structural component to the ground.

After the appropriate loads are calculated for the shelter, they should be applied to the exterior wall and roof surfaces of the shelter to determine the design forces for the structural and nonstructural elements. The continuous load path carries the loads acting on a building's exterior façade and roof through the floor diaphragms to the gravity load-bearing system and lateral load-bearing system. The individual components of the façade and roof must be able to develop these extraordinary forces, though

STRUCTURAL DESIGN CRITERIA

deformed, and transfer them to the underlying beams, trusses, girders, shear walls, and columns that provide the global structural resistance. These structural systems must also be able to develop uplift forces and load reversals that may accompany these extreme loading conditions. Uplift forces and load reversals are typically applied contrary to the conventional design loads and, therefore, details must be developed that account for these contrary patterns of deformation (see Figure 2-7). Seismic detailing that addresses ductile behavior despite multiple cycles of load reversals are generally well suited for all of these extreme loading conditions and building-specific details must consider each threat condition. Some construction materials, however, are better suited to developing a load path that can withstand loads from multiple directions and events. Cast-in-place reinforced concrete and steel moment frame construction is more commonly detailed to provide load paths than in "progressive collapse" designs utilizing panelized or masonry load-bearing construction. Nevertheless, appropriate details must be developed for nearly all structural systems.

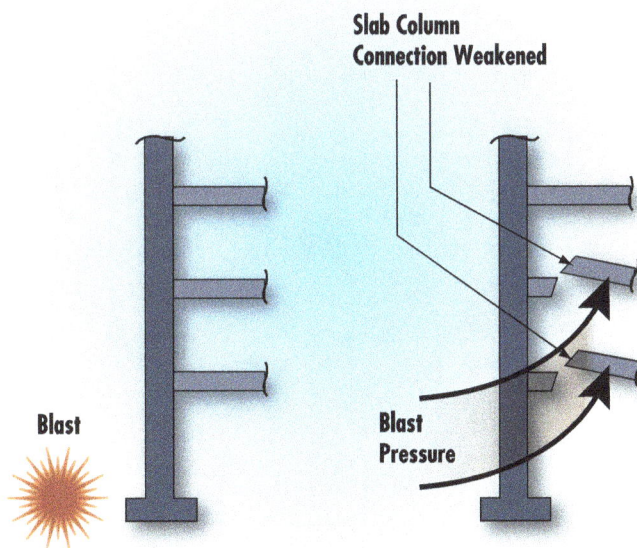

Figure 2-7
Effects of uplift and load reversals

Floor slabs are typically designed to resist downward gravity loading and have limited capacity to resist uplift pressures or the upward deformations experienced during load reversals that may precipitate a flexural or punching shear failure (see Figure 2-8). Therefore, floor slabs that may be subjected to significant uplift

Figure 2-8
Flat slab failure mechanisms

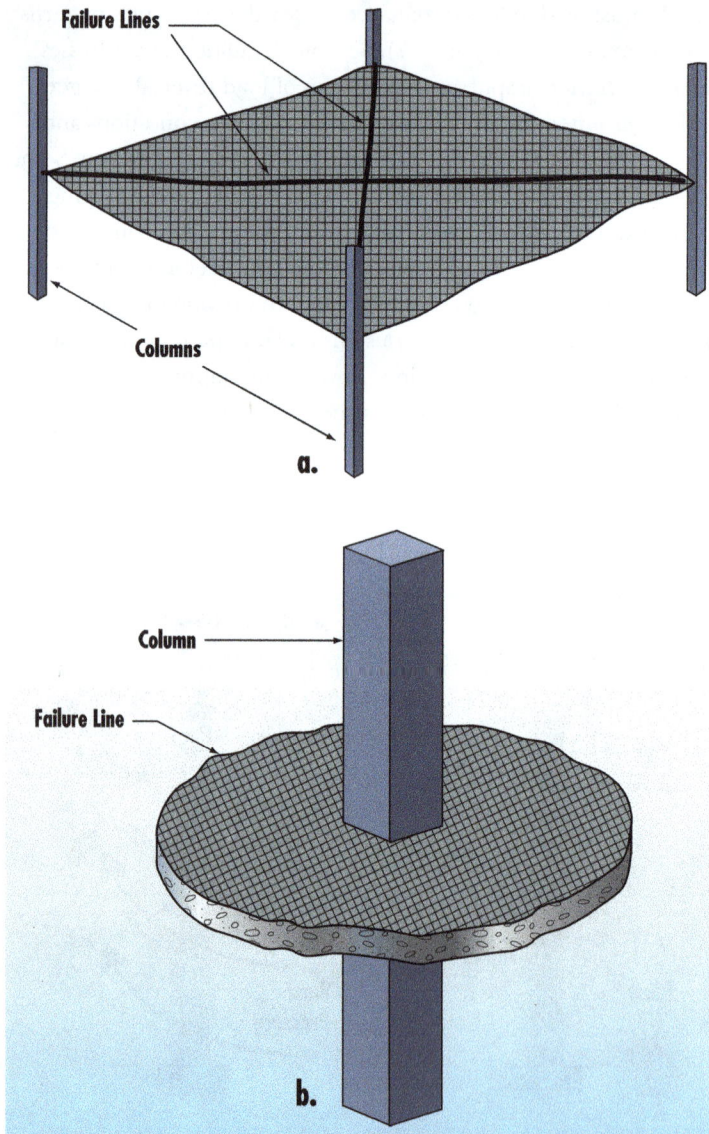

a.

b.

STRUCTURAL DESIGN CRITERIA

pressures, such that they overcome the gravity loads and subject the slabs to reversals in curvature, require additional reinforcement. If the slab does not contain this tension reinforcement, it must be supplemented with a lightweight carbon fiber application that may be bonded to the surface at the critical locations. Carbon fiber reinforcing mats bonded to the top surface of slabs would strengthen the floors for upward loading and reduce the likelihood of slab collapse from blast infill uplift pressures as well as internal explosions in mailrooms or other susceptible regions. This lightweight high tensile strength material supplements the limited capacity of the concrete to resist these unnatural loading conditions. An alternative approach would be to notch grooves in the top of concrete slabs and epoxy carbon fiber rods into grooves; although this approach may offer a greater capacity, it is much more invasive.

Similarly, adequate connections must be provided between the roof sheathing and roof structure to prevent uplift forces from lifting the roof off of its supports. Reinforcing steel, bolts, steel studs, welds, screws, and nails are used to connect the roof decking to the supporting structure. The detailing of these connections depends on the magnitude of the uplift or catenary forces that may be developed. The attachment of precast planks to the supporting structure will require special attention to the connection details. However, as with all other forms of construction, ductile and redundant detailing will produce superior performance in response to extreme loading.

Wall systems are typically connected to foundations using anchor bolts, reinforcing steel and imbedded plate systems properly welded together, and nailed mechanical fasteners for wood construction. Although these connections benefit from the weight of the structure bearing against the foundations and the lateral restraint provided by keyed details, the connections must be capable of developing the design forces in both the connectors and the materials into which the connectors are anchored.

2.3.3   Building Envelope

Façade components that must transfer the collected loads to the structural system must be designed and detailed to absorb significant amounts of energy associated with the extreme loading through controlled deformation. The duration of the extreme loading significantly influences the criteria governing the design of the building envelope systems. Significant inelastic deformations may be permitted for extraordinary events that impart the extreme loading over very short periods of time (e.g., explosive detonations). The building envelope system need only be designed to resist the direct shock wave, rebound, and any reflections off of neighboring buildings, all of which will occur within a matter of milliseconds (see Figure 2-9).

Figure 2-9
Blast damaged façade

Resistance to blast is often compared to resistance to natural hazards with the expectation that the protection against one will provide protection against the other. Therefore, as a first step, one should consider any inherent resistance derived from a building's design to resist environmental loading. Extreme wind loading resulting from tornadoes may similarly be of short enough duration to permit a large deformation of the façade in response to the peak loading. Certainly, the debris impact criteria will be similar to that for blast loading. However, hurricane winds may persist for extended periods of time and the performance criteria for façade components in response to these sustained pressures permit smaller deformations and less damage to the system. Breach of the façade components would permit pressures to fill the building and loads to be applied to nonstructural components. Anchorages and connections must be capable of holding the façade materials intact and attached to the building. Brittle modes of failure must be avoided to allow ductile deformations to occur.

2.3.4   Forced Entry and Ballistic Resistance

Ballistic-resistant design involves both the blocking of sightlines to conceal the occupants and the use of ballistic-resistant materials to minimize the effectiveness of the weapon. To reduce exposure, the safe room should be located as far as possible into the interior of the facility and walls should be arranged to eliminate sightlines through doorways. In order to provide the required level of resistance, the walls must be constructed using the appropriate thickness of ballistic-resistant materials, such as reinforced concrete, masonry, mild steel plate, or composite materials. The required thickness of these materials depends on the level of ballistic resistance; however, resistance to a high level of ballistic threat may be achieved using 6.5 inches of reinforced concrete, 8 inches of grouted concrete masonry unit (CMU) or brick, 1 inch mild steel plate, or ¾ inch armor steel plate. A ½-inch thick layer of bullet-resistant fiberglass may provide resistance up to a medium level of ballistic threat. Bullet-resistant doors are required for a high level of protection; however, hollow steel or steel clad doors with pressed steel frames may be used with an appropriate concealed entryway. Ballistic-resistant window assemblies contain multiple layers of laminated glass or polycarbonate materials and steel frames. Because these assemblies tend to be both heavy and expensive, their number and size should be minimized. Roof structures should contain materials similar to the ballistic-resistant wall assemblies. Ratings of bullet-resisting materials are presented in Table 2-1.

Table 2-1: UL 752 Ratings of Bullet-resisting Materials

Rating	Ammunition	Grain	Minimum Velocity (fps)
Level 1	9 mm full metal copper jacket with lead core	124	1,185
Level 2	.357 Magnum jacketed lead soft point	158	1,250
Level 3	.44 Magnum lead semi-wadcutter gas checked	240	1,350
Level 4	.30 caliber rifle lead core soft point	180	2,540
Level 5	7.62 mm rifle lead core full metal copper jacket, military ball	150	2,750
Level 6	9 mm full metal copper jacket with lead core	124	1,400
Level 7	5.56 mm rifle full metal copper jacket with lead core	55	3,080
Level 8	7.62 mm rifle lead core full metal copper jacket, military ball	150	2,750

UL = Underwriters Laboratories

Forced entry resistance is measured in the time it takes for an aggressor to penetrate the enclosure using a variety of hand tools and weapons. The required delay time is based on the probability of detecting the aggressors and the probability of a response force arriving within a specified amount of time. The different layers of defense create a succeeding number of security layers that are more difficult to penetrate, provide additional warning and response time, and allow building occupants to move into defensive positions or designated safe haven protection (see Figure 2-10). The rated delay time for each component comprising a defense layer (walls, doors, windows, roofs, floors, ceilings, and utility openings) must be known in order to determine the effective delay time for the safe room. Conventional construction offers little resistance to most forced entry threat severity levels and the rating of different forced entry-resistant materials is based on standardized testing under laboratory conditions.

STRUCTURAL DESIGN CRITERIA

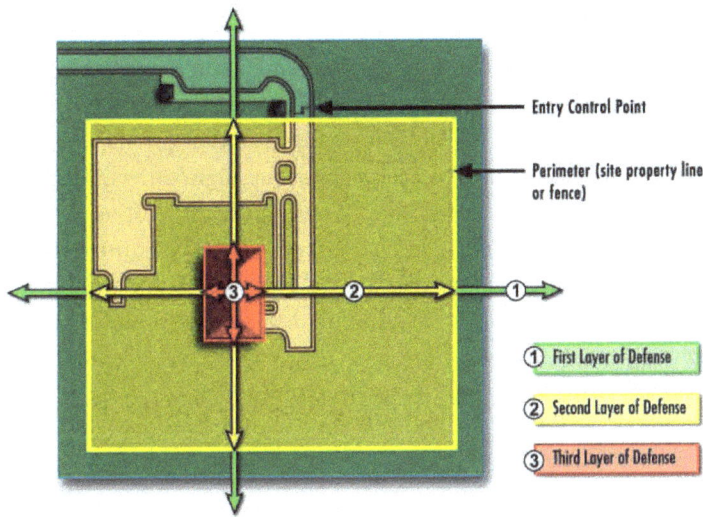

Figure 2-10
Layers of defense

Entry Control Point

Perimeter (site property line or fence)

① First Layer of Defense

② Second Layer of Defense

③ Third Layer of Defense

2.4 NEW CONSTRUCTION

The design of new buildings to contain shelters provides greater opportunities than the retrofit of existing buildings. Whether the entire building or just the shelter is to be resistant to the explosive terrorist threat may have a significant impact on the architectural and structural design of the building. Furthermore, unless the building is required to satisfy an established security design criteria, the weight of explosive that the building is to be designed to resist must be established by a site-specific threat and risk assessment. Even so, given the evolving nature of the terrorist threat, it is impossible to predict all the extreme conditions to which the building may be exposed over its life. Therefore, even if the building is not to be designed to resist any specific explosive threat, the American Society of Civil Engineers *Minimum Design Loads for Buildings and Other Structures* (ASCE-7) requires the building to be designed to sustain local damage without the building as a whole "being damaged to an extent disproportionate to the original local damage." The building can therefore be designed to prevent the progression of collapse in the unlikely event a primary member loses its load carrying capacity. This minimum design feature, achieved through the indirect prescriptive method or direct alternate path approach, will improve the structural integrity and provide an additional measure of safety to occupants. Incorporating continuity, redundancy, and ductility into the design will allow a damaged building to bridge over a failed element and redistribute loads through flexure or catenary action. This will limit the extent of debris that might otherwise rain down upon the hardened shelter. Where specific threats are defined, the vulnerable structural components may be hardened to withstand the intensity of explosive loading. The local hardening of vulnerable components in addition to the indirect prescriptive detailing of the structural system to bridge over damaged components will provide the most protection to the building.

2.4.1 Structure

Both steel frame and reinforced concrete buildings may be designed and detailed to resist the effects of an exterior vehicle explosive threat and an interior satchel explosion. Although steel construction may be more efficient for many types of loading, both conventional and unconventional, cast-in-place reinforced concrete construction provide an inherent continuity and mass that makes it desirable for blast-resistant buildings.

Reinforced concrete is a composite material in which the concrete provides the primary resistance to compression and shear and the steel reinforcement provides the resistance to tension and confines the concrete core. In addition to ductile detailing, which allows the reinforced concrete members to sustain large deformations and uncharacteristic reversals of curvature, the structural elements are typically stockier and more massive than their steel frame counterparts. The additional inertial resistance as well as the continuity of cast-in-place construction facilitates designs that are capable of sustaining the high intensity and short duration effects of close-in explosions. Furthermore, reinforced concrete buildings tend to crack and dissipate large amounts of energy through internal damping. This limits the extent of rebound forces and deformations.

Blast-resistant detailing requires continuous top and bottom reinforcement with tension lap splices staggered over the spans,
confinement of the plastic hinge regions by means of closely
spaced ties, and the prevention of shear failure prior to developing the flexural capacity (see Figure 2-11). One- or two-way
slabs supported on beams provide the best resistance to near contact satchel threats, which may produce localized breach, but allow
the structure to redistribute the gravity loads. Concrete columns
must be confined with closely spaced spiral ties, steel jackets, or
composite wraps. This confinement increases the shear resistance,
improves the ductility, and protects against the shattering effects
resulting from a near contact explosion. Cast-in-place exterior
walls or precast panels are best able to withstand a sizable stand-off
vehicular explosive threat and may be easily detailed to interact
with the reinforced concrete frame as part of the lateral load-resisting system.

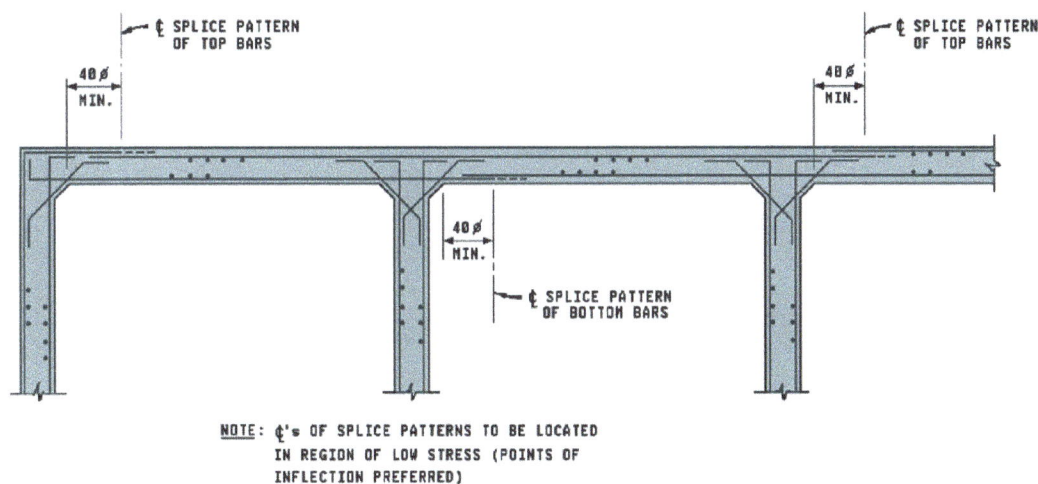

Figure 2-11   Multi-span slab splice locations

Steelwork is generally better suited to resist relatively low intensity, but long duration effects of large stand-off explosions. Steel is an inherently ductile material that is capable of sustaining large deformations; however, the very efficient thin-flanged sections make the conventional frame construction vulnerable to localized damage. Complex stress combinations and concentrations may occur that cause localized distress and prevent the section from developing its ultimate strength. Steel buildings may experience significant rebound and must therefore be designed to support significant reversals of loading. Concrete filled tube sections or concrete encased flanged sections may be used to protect the thin-flanged sections and supplement the inertial resistance. Concrete encasement should extend a minimum of 4 inches beyond the width and depth of the steel flanges and reinforcing bars may be detailed to tie into the concrete slabs.

To allow the concrete encasement to be tied into the floor slabs, the typical metal pan with concrete deck construction will require special detailing. Metal deck construction provides a spall shield to the underside of the slabs, which provides additional protection to a near contact satchel situated on a floor. However, the internal explosive threat will also load the ceiling slabs from beneath and the beams must contain an ample amount of studs, which far exceeds the requirements for conventional gravity design, to transfer the slab reactions to the steel supports without pulling out. If the slabs are adequately connected to the steel-framing members, these beams will be subjected to abnormal reversals of curvature. These reversals will subject the mid-span bottom flanges to transient compressive stress and may induce a localized buckling. Because the blast loads are transient, the dominant gravity loads will eventually restore the mid-span bottom flange to tension; however, unless it is adequately braced, the transient buckling will produce localized damage.

The concrete encasement of the steel beams will provide torsional resistance to the cross-section and minimize the need for intermediate bracing. If the depth of the composite section is to be minimized by embedding the steel section into the thickness of the slab, the slab reinforcement must either be welded to the webs or run through holes drilled into the webs in order to maintain continuity. All welding of reinforcing steel must be in accordance with seismic detailing to prevent brittle failures. Steel columns require full moment splices and the relatively thin flange sections require concrete encasement to prevent localized damage. To take full advantage of the steel capacity and dissipate the greatest amount of energy through ductile inelastic deformation, the beam to column connections must be capable of developing the plastic flexural capacity of the members. Connection details, similar to those used in seismic regions, will be required to develop the corresponding flexural and shear capacity (see Figure 2-12). Connecting exterior cast-in-place reinforced concrete walls to the steel frame will require details that transfer both the direct blast loads in bearing and the subsequent rebound effects in tension. Precast panels are simply supported at the ends and, unless they span over multiple floors, they lack the continuity of monolithic cast-in-place wall construction. Cold joints in the cast-in-place construction require special detailing and the connection details for the precast panels must be able to resist both the direct blast loads in bearing and the subsequent rebound effects in tension.

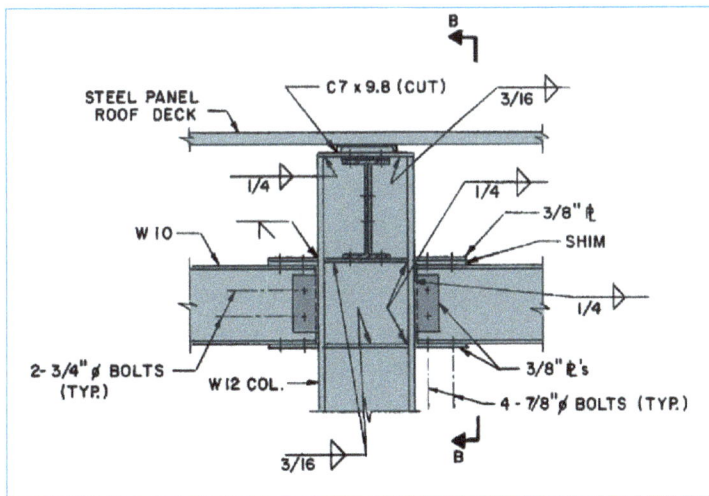

Figure 2-12
Typical frame detail at interior column

Regardless of the materials, framed buildings perform best when column spacing is limited and the use of transfer girders is limited. Bearing wall systems that rely on interior cross-walls will benefit from periodically spaced longitudinal walls that enhance stability and control the lateral progression of damage. Bearing wall systems that rely on exterior walls will benefit from periodically spaced perpendicular walls or substantial pilasters that limit the extent of wall that is likely to be affected.

Free-standing columns do not have much surface area; therefore, air-blast loads on columns are limited by clear-time effects in which relief waves from the free edges attenuate the reflected intensity of the blast loads. Where the exterior façade inhibits clear-time effects prior to façade failure, the columns will receive the full intensity of the reflected blast pressures. Large stand-off explosive threats may produce large inelastic flexural deformations that could initiate P-Δ induced instabilities. Short stand-off explosive threats may cause shear, base plate, or column splice failures. Near contact threats may cause brisance, which is the shattering of reinforced concrete sections. Confinement of reinforced concrete members by means of spiral reinforcement, steel jackets, or carbon fiber wraps may improve their resistance. Encasement of steel sections will inhibit local flange and web plate deformations that could precipitate a section failure. Exterior column splices should be located as high above grade level as practical and match the capacity of the column section.

Load-bearing walls do not benefit from clear-time effects as columns do and therefore collect the full intensity of the reflected blast pressure pulse. Nevertheless, reinforced concrete load-bearing walls are particularly effective if adequately reinforced. Fully grouted masonry walls, on the other hand, are more brittle and seismic levels of reinforcement greatly increase the ductility and performance of masonry walls. Continuous reinforced bond beams, with a minimum of one #4 bar or equivalent, are required in the wall at the top and bottom of each floor and roof level. Interior horizontal ties are required in the floors perpendicular to the wall. The ties are equivalent to a #4 bent bar at a maximum spacing of 16 inches that extends into the slab and the wall the greater of the development length of the bar or 30 inches. Vertical ties are required from floor to floor at columns, piers, and walls. The ties should be equivalent to a #4 bar at a maximum spacing of 16 inches coinciding with the horizontal ties. The ties should be continuous through the floor and extend into the wall above and below the floor the greater of the development length of the bar or 30 inches. Partition walls surrounding critical systems or isolating areas of internal threat, such as lobbies, loading docks, and mailrooms, require fully grouted reinforced masonry construction. It is particularly difficult to extend the reinforcement to the full height of the partition wall and develop the reaction forces. Reinforced bond beams are required as for load-bearing walls.

Flat roof systems are exposed to the incident blast pressures that diffuse over the top of the building, causing complex patterns of shadowing and focusing on the surface. Subsequent negative phase effects may subject the pre-weakened roof systems to low intensity, but long duration suction pressures; therefore, lightweight roof systems may be susceptible to uplift effects. Two-way beam slab systems are preferred for reinforced concrete construction and metal deck with reinforced concrete fill is preferred for steel frame construction. Both of these roof systems provide the required mass, strength, and continuity to resist all phases of blast loading. The performance of conventional precast concrete plank systems depends to a great extent on the connection details, and these connections need to be detailed to provide continuity. Flat slab and flat plate construction requires continuous bottom reinforcement in both directions to improve the integrity and special details at the columns to prevent a punching shear failure. Post-tensioned slab systems are particularly problematic because the cable profile is typically designed to resist the predominant patterns of gravity load and the system is inherently weak in response to load reversals.

2.4.2   Façade and Internal Partitions

The building's façade is its first real defense against the effects of a bomb and is typically the weakest component that will be subjected to blast pressures. Debris mitigating façade systems may be designed to provide a reasonable level of protection to a low or moderate intensity threat; however, façade materials may be locally overwhelmed in response to a low intensity short stand-off detonation or globally overwhelmed in response to a large intensity long stand-off detonation. As a result, it is unreasonable to design a façade to resist the actual pressures resulting from the design level threat everywhere over the surface of the building. In fact, successful performance of the blast-resistant façade may be defined as throwing debris with less than high hazard velocities. This is particularly true for the glazed fenestration. The peak pressures and impulses that are used to select the laminated glazing makeup are typically established such that no more than 10 percent of the glazed fenestration will produce debris that is propelled with high hazard velocities into the occupied space in response to any single detonation of the design level threat. The definitions of high hazard velocities were adapted from the United Kingdom hazard guides and correspond to debris that is propelled 10 feet from the plane of the glazing and strikes a witness panel higher than 2 feet above the floor. Similarly, a medium level of hazard corresponds to debris that strikes the witness panel no higher than 2 feet above the floor. A low level of hazard corresponds to debris that strikes the floor no farther than 10 feet from the plane of the glazing and a very low level of hazard corresponds to debris that strikes the floor no farther than 3.3 feet from the plane of the glazing. Glass hazard response software was developed for the U.S. Army Corps of Engineers, the General Services Administration, and the Department of State to determine the performance of a wide variety of glazing systems in response to blast loading. These simplified single-degree-of-freedom dynamic analyses account for the strength of the glass prior to cracking and the post-damage capacity of the laminated interlayers. While many of these glass hazard response software remain restricted, the American Society for Testing and Materials (ASTM) 2248 relates the design of glass to resist blast loading to an equivalent 3-second equivalent wind load.

In order for the glazing to realize its theoretical capacity, it must be retained by the mullions with an adequately sized bite, by means of a structural silicone adhesive, or a combination of the two. Furthermore, in order for the mechanical bite and silicone adhesive to be effective, the mullion deformations over the length of the lite must be limited (see Figure 2-13). Unfortunately, the maximum extent of deformation that the mullion may sustain prior to dislodging the glass is poorly defined. A conservative limit of 2 degrees is often assumed for typical protective glazing systems; however, advanced analytics may justify a significantly greater mullion deformation limit. Mullions must therefore be able to accept the reaction forces from the edges of the glazed elements and remain intact and attached to the building. Analyses of mullion deformations and anchorage details are required to demonstrate the safe performance of the glazed fenestration.

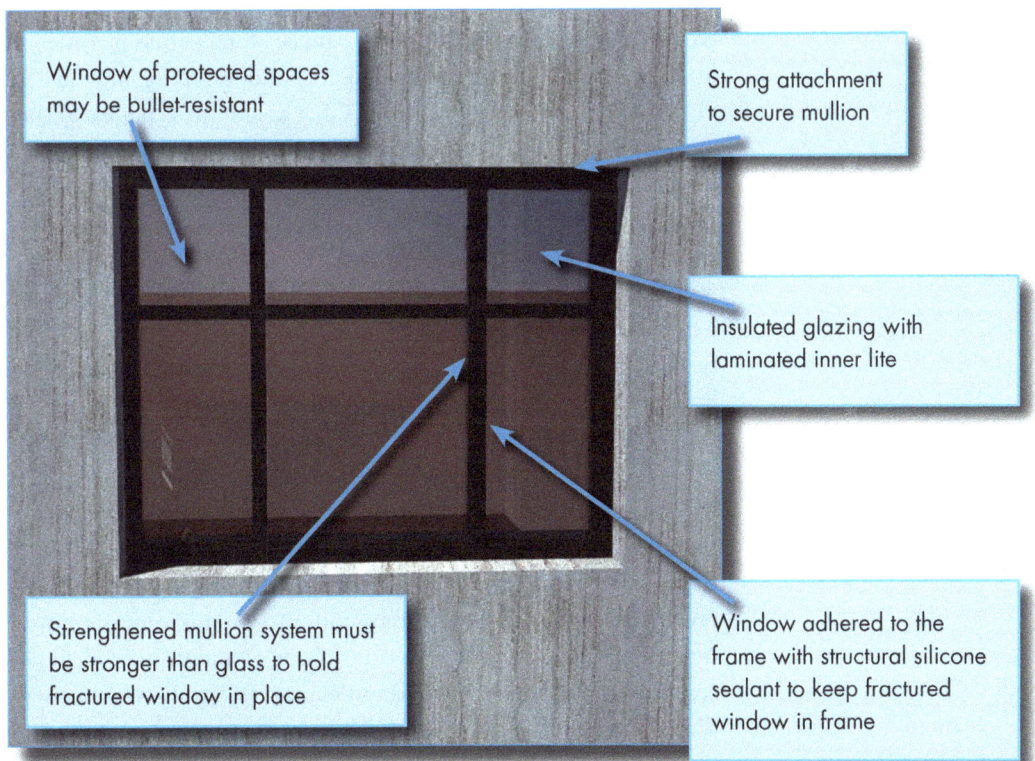

Figure 2-13   Protective façade design

Curtainwall systems are inherently lightweight and flexible façade systems; however, well designed curtainwall systems demonstrated, through explosive testing, considerable resilience in response to blast loading. Furthermore, the glazed components are subjected to less intense loads as their flexible supports deform in response to the blast pressures. A multi-degree-of-freedom model of the façade will determine the accurate interaction of the individual mullions and the phasing of the interconnecting forces. Because all response calculations must be dynamic and inelastic, the accurate representation of the phasing of these forces may significantly affect the performance. Curtainwall anchors are attached directly to the floor slabs where the large lateral loads may be transferred directly through the diaphragms into the lateral load-resisting systems.

Façade systems may contain combinations of glazing, metal panels, precast concrete, or stone panels. Metal panels provide little inertial resistance, but are capable of developing large inelastic deformations. The fasteners that attach these panels to the mullions or metal studs must be designed to transfer the large membrane forces. Stone panels provide significant inertial resistance, but are relatively brittle and have little strength beyond their modulus of rupture. Stud wall systems that restrain these façade panels may deform within acceptable levels and develop a membrane stiffening capacity, and strain energy methods may be used to calculate their response. However, the anchorage of the studs to the floor and ceiling slabs are likely to limit the forces they can develop.

Precast panels may easily be designed to provide inelastic deformation in response to the design level threats. However, the design of their anchorage to hold them to the building during both the direct loading and subsequent rebound phase require more robust details. Because the primary load carrying elements may buckle in response to the large collected forces, precast panels are attached directly to the floor slabs where the forces may be transferred through the diaphragms to the lateral load-resisting elements. Where mullions are attached within punched out openings in precast panels, the spacing of the anchorages will determine the span of the mullions and the force each anchorage is required to resist. Embedded anchors within the precast panels will be required to accept these anchorage forces.

Fully grouted and reinforced CMU façades may be designed to accept the large lateral loads produced by blast events; however, it is often difficult to detail them to transfer the reaction forces to the floor slabs. A continuous exterior CMU wall that bears against the floor slabs may avoid many of the construction and connection difficulties, but this is not typical construction practice. Brick or stone veneer does not appreciably increase the strength of the CMU wall, but the added mass increases its inertial resistance.

2.5 EXISTING CONSTRUCTION: RETROFITTING CONSIDERATIONS

Although retrofitting existing buildings to include a shelter can be expensive and disruptive to users, it may be the only available option. When retrofitting existing space within a building is considered, data centers, interior conference rooms, stairwells, and other areas that can be structurally and mechanically isolated provide the best options. Designers should be aware that an area of a building currently used for refuge may not necessarily be the best candidate for retrofitting when the goal is to provide comprehensive protection.

An existing area that has been retrofitted to serve as a shelter is unlikely to provide the same degree of protection as a shelter designed as new construction. When existing space is retrofitted for shelter use, issues have arisen that have challenged both designers and shelter operators. For example, glass and unreinforced masonry façades are particularly vulnerable to blast loading. Substantial stand-off distances are required for the unprotected structure and these distances may be significantly reduced through the use of debris mitigating retrofit systems. Furthermore, because blast loads diminish with distance and incidence of blast wave to the loaded surface, the larger threats at larger stand-off distances are likely to damage a larger percentage of façade elements than the more localized effects of smaller threats at shorter stand-off distances. Safe rooms that may be located within a building should therefore be located in windowless spaces or spaces in which the window glazing was upgraded with a fragment retention film (FRF).

2.5.1 Structure

The building's lateral load-resisting system, the structural frame or shear walls that resist wind and seismic loads, will be required to receive the blast loads that are applied to the exterior façade and transfer them to the building's foundation. This load path is typically through the floor slabs that act as diaphragms and interconnect the different lateral load-resisting elements. The lateral load-resisting system for a building depends to a great extent on the type of construction and region. In many cases, low-rise buildings do not receive substantial wind and seismic forces and, therefore, do not require substantial lateral load-resisting systems. Because blast loads diminish with distance, a package sized explosive threat is likely to locally overwhelm the façade, thereby limiting the force that may be transferred to the lateral load-resisting system. However, the intensity of the blast loads that may be applied to the building could exceed the design limits for most conventional construction. As a result, the building is likely to be subjected to large inelastic deformations that may produce severe cracks to the structural and nonstructural partitions. There is little that can be done to upgrade the existing structure to make it more ductile in response to a blast loading that doesn't require extensive renovation of the building; therefore, safe rooms should be located close to the interior shear walls or reinforced masonry walls in order to provide maximum structural support in response to these uncharacteristically large lateral loads.

Unless the structure is designed to resist an extreme loading, such as a hurricane or an earthquake, it is not likely to sustain extensive structural damage without precipitating a progressive collapse. The effects of a satchel-sized explosive in close contact to a column or a vehicle-borne explosive device at a sidewalk's distance from the façade may initiate a failure of a primary structure that may propagate as the supported loads attempt to redistribute to an adjoining structure. Transfer girders that create long span structures and support large tributary areas are particularly susceptible to localized damage conditions. As a result, safe rooms should not be located on a structure that is either supported by or underneath a structure that is supported by transfer girders unless the building is evaluated by a licensed professional engineer. The connection details for multi-story precast structures should also be evaluated before the building is used to house a safe room.

Nonstructural building components, such as piping, ducts, lighting units, and conduits that are located within safe rooms must be sufficiently tied back to a solid structure to prevent failure of the services and the hazard of falling debris. To mitigate the effects of in-structure shock that may result from the infilling of blast pressures through damaged windows, the nonstructural systems should be located below the raised floors or tied to the ceiling slabs with seismic restraints.

2.5.2   Façade and Internal Partitions

Safe rooms in existing buildings should be selected to provide the space required to accommodate the building population and should be centrally located to allow quick access from any location within the building, enclosed with fragment mitigating partitions or façade, and within robust structural systems that will resist collapse. These large spaces are best located at the lower floors, away from a lightweight roof and exterior glazing elements. If such a space does not exist within the existing building, the available spaces may be upgraded to achieve as many of these attributes as possible. This will involve the treatment of the exterior façade with fragment mitigating films, blast curtains, debris catch systems, spray-on applications of elasto-polymers to unreinforced masonry walls, and hardening of select columns and slabs with composite fiber wraps, steel jackets, or concrete encasements.

2.5.2.1 Anti-shatter Façade. The conversion of existing construction to provide blast-resistant protection requires upgrades to the most fragile or brittle elements enclosing the safe room. Failure of the glazed portion of the façade represents the greatest hazard to the occupants. Therefore, the exterior glazed elements of the façade and, in particular, the glazed elements of the designated safe rooms, should be protected with an FRF, also commonly known as anti-shatter film (ASF), "shatter-resistant window film" (SRWF), or "security film." These materials consist of a laminate that will improve post-damage performance of existing windows. Applied to the interior face of glass, ASF holds the fragments of broken glass together in one sheet, thus reducing the projectile hazard of flying glass fragments.

Most ASFs are made from polyester-based materials and coated with adhesives. ASFs are available as clear, with minimal effects to the optical characteristics of the glass, and tinted, which provides a variety of aesthetic and optical enhancements and can increase the effectiveness of existing heating/cooling systems. Most films are designed with solar inhibitors to screen out ultraviolet (UV) rays and are available treated with an abrasion-resistant coating that can prolong the life of tempered glass.[1] However, over time, the UV absorption damages the film and degrades its effectiveness.

According to published reports, testing has shown that a 7-mil thick film, or specially manufactured 4-mil thick film, is the minimum thickness that is required to provide hazard mitigation from blast. Therefore, a 4-mil thick ASF should be utilized only if it has demonstrated, through explosive testing, that it is capable of providing the desired hazard level response.

The application of security film must, at a minimum, cover the clear area of the window. The clear area is defined as the portion of the glass unobstructed by the frame. This minimum application, termed daylight installation, is commonly used for retrofitting windows. By this method, the film is applied to the exposed glass

---

[1] Abrasions on the faces of tempered glass reduce the glass strength.

without any means of attachment or capture within the frame. Application of the film to the edge of the glass panel, thereby extending the film to cover the glass within the bite, is called an edge to edge installation and is often used in dry glazing installations. Other methods of retrofit application may improve the film performance, thereby reducing the hazards; however, these are typically more expensive to install, especially in a retrofit situation.

Although a film may be effective in keeping glass fragments together, it may not be particularly effective in retaining the glass in the frame. ASF is most effective when it is used with a blast tested anchorage system. Such a system prevents the failed glass from exiting the frame (see Figure 2-14).

The wet glazed installation, a system where the film is positively attached to the frame, offers more protection than the daylight installation. This system of attaching the film to the frame reduces glass fragmentation entering the building. The wet glazing system utilizes a high strength liquid sealant, such as silicone, to attach the glazing system to the frame. This method is more costly than the daylight installation.

Securing the film to the frame with a mechanically connected anchorage system further reduces the likelihood of the glazing system exiting the frame. Mechanical attachment includes anchoring methods that employ screws and/or batten strips that anchor the film to the frame along two or four sides. The mechanical attachment method can be less aesthetically pleasing when compared to wet glazing because additional framework is necessary and is more expensive than the wet glazed installation.

Window framing systems and their anchorage must be capable of transferring the blast loads to the surrounding walls. Unless the frames and anchorages are competent, the effectiveness of the attached films will be limited. Similarly, the walls must be able to withstand the blast loads that are directly applied to them and accept the blast loads that are transferred by the windows. The strength of these walls may limit the effectiveness of the glazing upgrades.

Figure 2-14
Mechanically attached
anti-shatter film

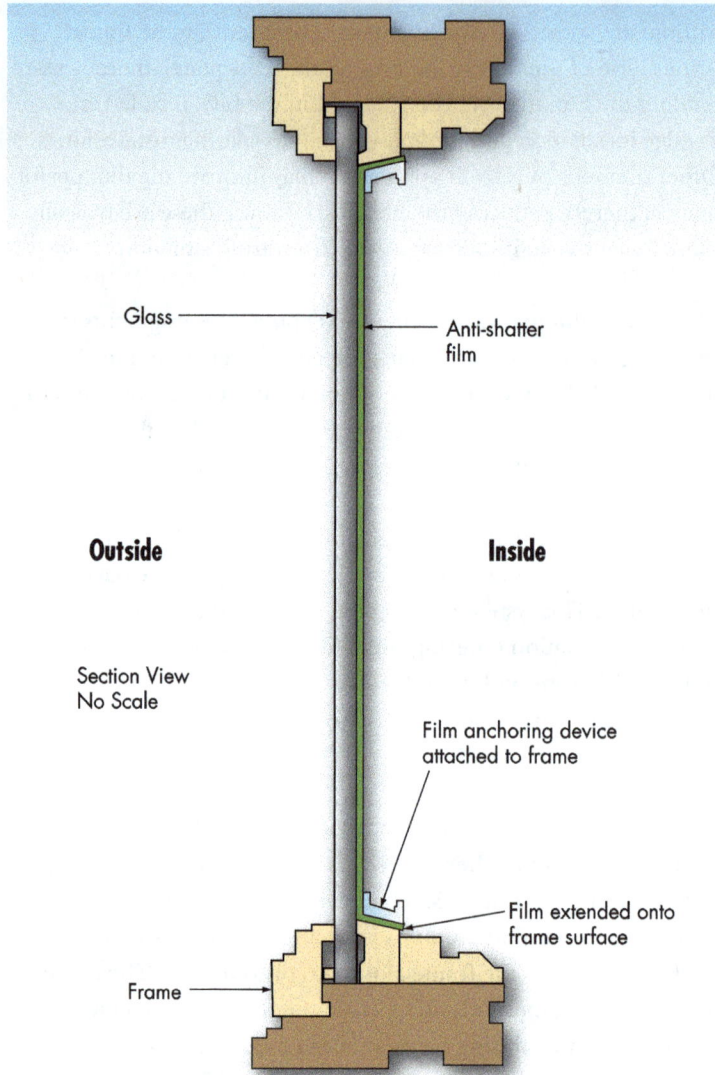

Figure 2-14
Mechanically attached
anti-shatter film

Glass

Anti-shatter
film

Outside

Inside

Section View
No Scale

Film anchoring device
attached to frame

Film extended onto
frame surface

Frame

If a major rehabilitation of the façade is required to improve the
mechanical characteristics of the building envelope, a laminated
glazing replacement is recommended. Laminated glass consists
of two or more pieces of glass permanently bonded together by
a tough plastic interlayer made of polyvinyl butyral (PVB) resin.
Once sealed together, the glass "sandwich" behaves as a single
unit. Annealed, heat strengthened, tempered glass, or

polycarbonate glazing can be mixed and matched between layers of laminated glass in order to design the most effective lite for a given application. When fractured, fragments of laminated glass tend to adhere to the PVB interlayer rather than falling free and potentially causing injury.

Laminated glass can be expected to last as long as ordinary glass, provided it is not broken or damaged in any way. It is very important that laminated glass is correctly installed to ensure long life. Regardless of the degree of protection required from the window, laminated glass needs to be installed with adequate sealant to prevent water from coming in contact with the edges of the glass. A structural sealant will adhere the glazing to the frame and allow the PVB interlayer to develop its full membrane capacity. Similar to attached film upgrades, the window frames and anchorages must be capable of transferring the blast loads to the surrounding walls.

2.5.2.2 Façade Debris Catch Systems. Blast curtains are made from a variety of materials, including a warp knit fabric or a polyethylene fiber. The fiber can be woven into a panel as thin as 0.029 inch that weighs less than 1.5 ounces per square foot. This fact dispels the myth that blast curtains are heavy sheets of lead that completely obstruct a window opening and eliminate all natural light from the interior of a protected building. The blast curtains are affixed to the interior frame of a window opening and essentially catch the glass fragments produced by a blast wave. The debris is then deposited on the floor at the base of the window. Therefore, the use of these curtains does not eliminate the possibility of glass fragments penetrating the interior of the occupied space, but instead limits the travel distance of the airborne debris. Overall, the hazard level to occupants is significantly reduced by the implementation of the blast curtains. However, a person sitting directly adjacent to a window outfitted with a blast curtain may still be injured by shards of glass in the event of an explosion.

The main components of any blast curtain system are the curtain itself, the attachment mechanism by which the curtain is affixed to the window frame, and either a trough or other retaining mechanism at the base of the window to hold the excess curtain material. The blast curtain with curtain rod attachment and sill trough differ largely from one manufacturer to the next. The curtain fabric, material properties, method of attachment, and manner in which they operate all vary, thereby providing many options within the overall classification of blast curtains. This fact makes blast curtains applicable in many situations.

Blast curtains differ from standard curtains in that they do not open and close in the typical manner. Although blast curtains are intended to remain in a closed position at all times, they may be pulled away from the window to allow for cleaning and blind or shade operation. However, the curtains can be rendered ineffective if installed such that easy access would provide opportunity for occupants to defeat their operation. The color and openness factor of the fabric contributes to the amount of light that is transmitted through the curtains and the see-through visibility of the curtains. Although the color and weave of these curtains may be varied to suit the aesthetics of the interior décor, the appearance of the windows is altered by the presence of the curtains.

The curtains may either be anchored at the top and bottom of the window frame or anchored at the top only and outfitted with a weighted hem. The curtain needs to be extra long, with the surplus either wound around a dynamic tension retainer or stored in a reservoir housing. When an explosion occurs, the curtain feeds out of the receptacle to absorb the force of the flying glass fragments. The effectiveness of the blast curtains relies on their use and no protection is provided when these curtains are pulled away from the glazing (see Figure 2-15).

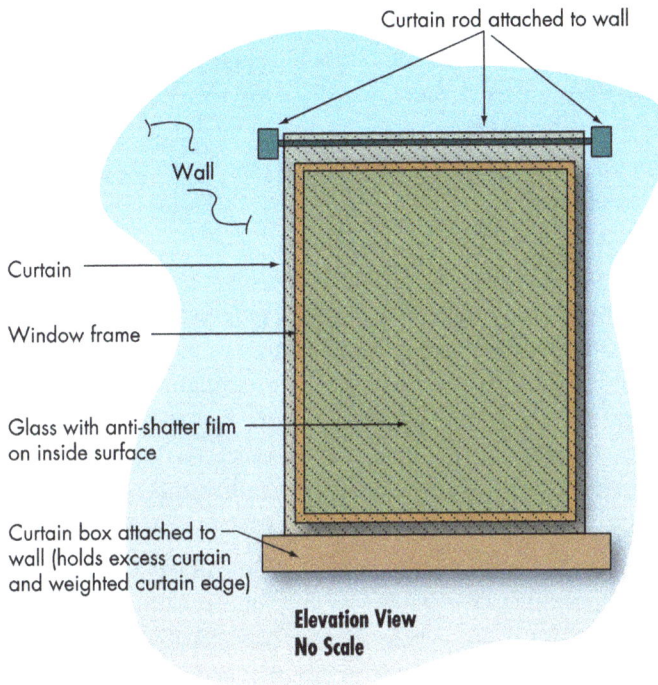

Curtain rod attached to wall

Figure 2-15
Blast curtain system

Wall

Curtain

Window frame

Glass with anti-shatter film
on inside surface

Curtain box attached to
wall (holds excess curtain
and weighted curtain edge)

**Elevation View
No Scale**

Rigid catch bar systems were designed and tested as a means of increasing the effectiveness of filmed and laminated window up-grades. Anti-shatter film and laminated glazing are designed to hold the glass shards together as the window is damaged; however, unless the window frames and attachments are upgraded as well to withstand the capacity of the glazing, this retrofit will not prevent the entire sheet from flying free of the window frames. The rigid catch bars intercept the filmed or laminated glass and disrupt their flight; however, they are limited in their effectiveness, tending to break the dislodged façade materials into smaller projectiles.

Rigid catch systems collect huge forces upon impact and require considerable anchorage into a very substantial structure to prevent failure. If either the attachments or the supporting structure are incapable of restraining the forces, the catch system will be dislodged and become part of the debris. Alternatively, the debris may be sliced by the rigid impact and the effectiveness of the catch bar will be severely reduced. Finally, the effectiveness of debris catch systems are limited where double pane, insulated glazing units (IGUs) are used. Since anti-shatter film or laminated glass is typically applied to only the inner surface of an IGU, debris from the damaged outer lite could be blown past the catch bar into the protected space.

Flexible catch bars can be designed to absorb a significant amount of the energy upon impact, thereby keeping the debris intact and impeding their flight. These systems may be designed to effectively repel the debris and inhibit their flight into the occupied spaces; they also may be designed to repel the debris from the failed glazing as well as the walls in which the windows are mounted. The design of the debris restraint system must be strong enough to withstand the momentum transferred upon impact and the connections must be capable of transferring the forces to the supporting slabs and spandrel beams. However, under no circumstances can the design of the restraint system add significant amounts of mass to the structure that may be dislodged and present an even greater risk to the occupants of the building.

Cables are extensively used to absorb significant amounts of energy upon impact and their flexibility makes them easily adaptable to many situations. The diameter of the cable, the spacing of the strands, and the means of attachment are all critical in designing an effective catch system. These catch cable concepts have been used by protective design window manufacturers as restraints for laminated lites. The use of cable systems has long been recognized as an effective means of stopping massive objects moving at high velocity. An analytical simulation or a physical test is required to confirm the adequacy of the cable catch system to restrain the debris resulting from an explosive event.

High performance energy absorbing cable catcher systems retain glass and frame fragments and limit the force transmitted to the supporting structure. These commercially available retrofit products consist of a series of ¼-inch diameter stainless steel cables connected with a shock-absorbing device to an aluminum box section, which is attached to the jambs, the underside of the header, and topside of the sill. The energy absorbing characteristics allow the catch systems to be attached to relatively weakly constructed walls without the need for additional costly structural reinforcement. To reduce the possibility of slicing the laminated glass, the cable may either be sheathed in a tube or an aluminum strip may be affixed to the glass directly behind the cable.

2.5.2.3 Internal Partitions. Unreinforced masonry walls provide limited protection against airblast due to explosions. When subjected to overload from air blast, brittle unreinforced CMU walls will fail and the debris will be propelled into the interior of the structure, possibly causing severe injury or death to the occupants. This wall type has been prohibited for new construction where protection against explosive threats is required. Existing unreinforced CMU walls may be retrofitted with a sprayed-on polymer coating to improve their air blast resistance. This innovative retrofit technique takes advantage of the toughness and resiliency of modern polymer materials to effectively deform and dissipate the blast energy while containing the shattered wall fragments. Although the sprayed walls may shatter in a blast event, the elastomer material remains intact and contains the debris.

The blast mitigation retrofit for unreinforced CMU walls consists of an interior and optional exterior layer of polyurea applied to exterior walls and ceilings (see Figure 2-16). The polyurea provides a ductile and resilient membrane that catches and retains secondary fragmentation from the existing concrete block as it breaks apart in response to an air blast wave. These fragments, if allowed to enter the occupied space, are capable of producing serious injury or death to occupants of the structure.

Figure 2-16
Spray-on elastomer coating

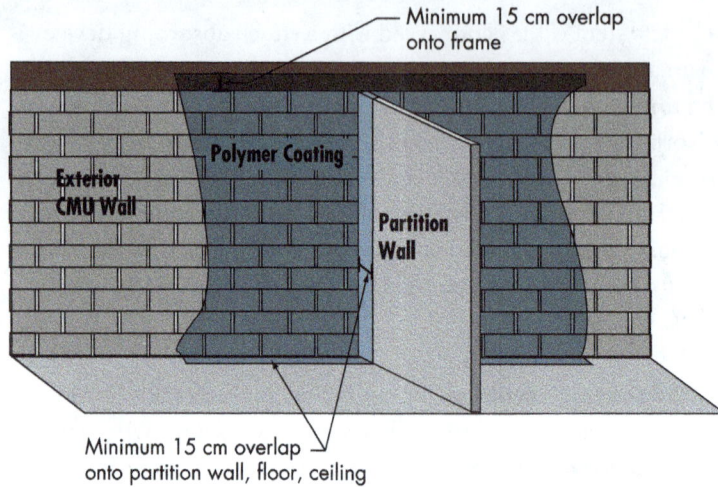

Minimum 15 cm overlap
onto frame

Polymer Coating

Exterior
CMU Wall

Partition
Wall

Minimum 15 cm overlap
onto partition wall, floor, ceiling

In lieu of the elastomer, an aramid (Geotextile) debris catching system may be attached to the structure by means of plates bolted through the floor and ceiling slabs (see Figure 2-17). Similar to the elastomer retrofit, the aramid layer does not strengthen the wall; instead, it restrains the debris that would otherwise be propelled into the occupied spaces.

Figure 2-17
Geotextile debris catch
system

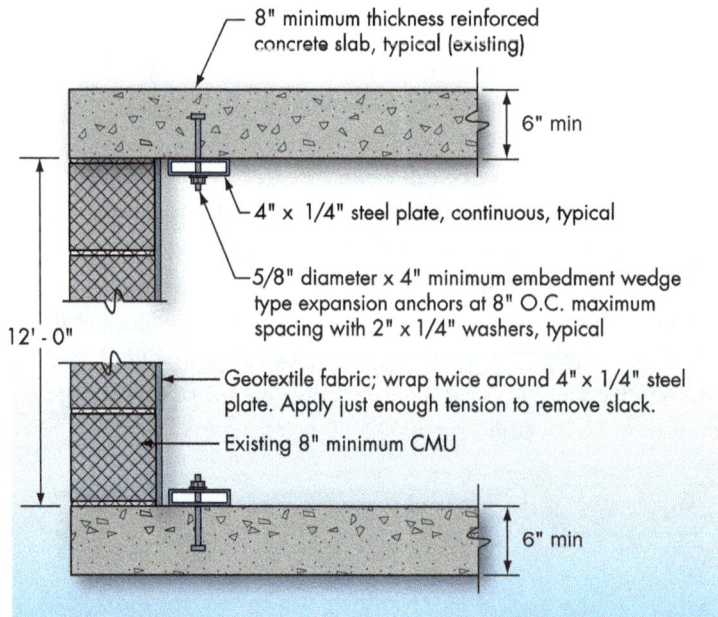

8" minimum thickness reinforced
concrete slab, typical (existing)

6" min

4" x 1/4" steel plate, continuous, typical

5/8" diameter x 4" minimum embedment wedge
type expansion anchors at 8" O.C. maximum
spacing with 2" x 1/4" washers, typical

12' - 0"

Geotextile fabric; wrap twice around 4" x 1/4" steel
plate. Apply just enough tension to remove slack.

Existing 8" minimum CMU

6" min

STRUCTURAL DESIGN CRITERIA

Alternatively, an unreinforced masonry wall may be upgraded with an application of shotcrete sprayed onto the wall with a welded wire fabric. This method supplements the tensile capacity of the existing wall and limits the extent of debris that might be propelled into the protected space. Steel sections may also be installed up against existing walls to reduce the span and provide an alternate load transfer to the floor diaphragms. Load-bearing masonry walls require additional redundancy to prevent the initiation of a catastrophic progression of collapse. Therefore, the fragment protection that may be provided by a spray-on elasto-polymer, a fabric spall shield, or a metal panel must be supplemented with structural supports that can sustain the gravity loads in the event of excessive wall deformation. The design of stiffened steel-plate wall systems to withstand the effects of explosive loading is one way of achieving such redundancy and fragment protection. These load-bearing wall retrofits require a more stringent design, capable of resisting lateral loads and the transfer of axial forces. Stiffened wall panels, consisting of steel plates to catch the debris and welded tube sections spaced some 3 feet on center to supplement the gravity load carrying capacity of the bearing walls, must be connected to the existing floor and ceiling slabs by means of base plates and anchor bolt connectors (see Figure 2-18).

A steel stud wall construction technique may also be used for new buildings or the retrofit of existing structures requiring blast resistance. Commercially available 18-gauge steel studs may be attached web to web (back to back) and 16-gauge sheet metal may be installed outboard of the steel studs behind the cladding (see Figure 2-19). While the wall absorbs a considerable amount of the blast energy through deformation, its connection to the surrounding structure must develop the large tensile reaction forces. In order to prevent a premature failure, these connections should be able to develop the ultimate capacity of the stud in tension. Ballistic testing of various building cladding materials requires a nominal 4-inch thickness of stone, brick, masonry, or concrete. Forced entry protection requires a ¼-inch thick layer of A36 steel plate that is behind the building's veneer and welded or screwed to the steel stud framing in lieu of the 16-gauge sheet metal.

Figure 2-18
Stiffened wall panels

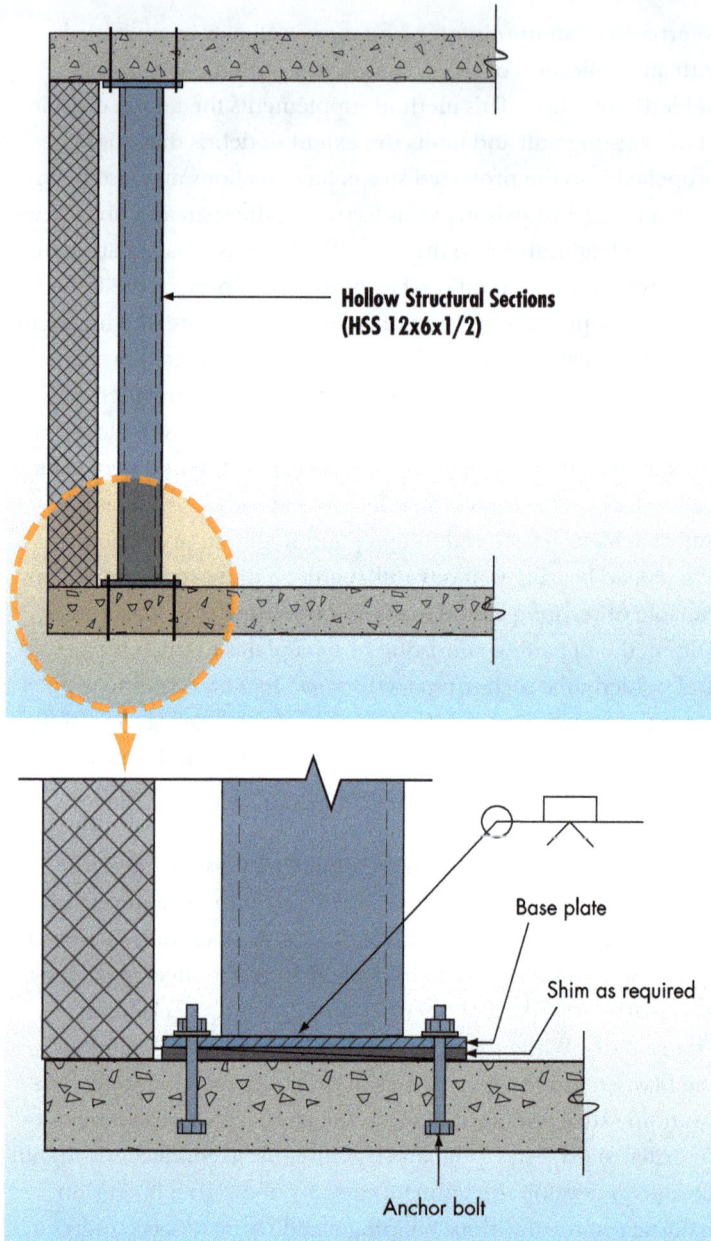

**Hollow Structural Sections
(HSS 12x6x1/2)**

Base plate

Shim as required

Anchor bolt

Figure 2-19   Metal stud blast wall

Internal installations require an interstitial sheathing of ½-inch A36 steel plate. Regardless whether a ¼-inch steel plate or a 16-gauge sheet metal is used, the interior face of the stud should be finished with a steel-backed composite gypsum board product.

2.5.2.4 Structural Upgrades. Conventionally designed columns may be vulnerable to the effects of explosives, particularly when placed in contact with their surface. Stand-off elements, in the form of partitions and enclosures, may be designed to guarantee a minimum stand-off distance; however, this alone may not be sufficient. Additional resistance may be provided to reinforced concrete structures by means of a steel jacket or a carbon fiber wrap that effectively confines the concrete core, thereby increasing the confined strength and shear capacity of the column, and holds the rubble together to permit it to continue carrying the axial loads (see Figure 2-20). The capacity of steel flanged columns may be increased with a reinforced concrete encasement that adds mass to the steel section and protects the relatively thin flange sections. The details for these retrofits must be designed to resist the specific weight of explosives and stand-off distance.

Figure 2-20
Steel jacket retrofit detail

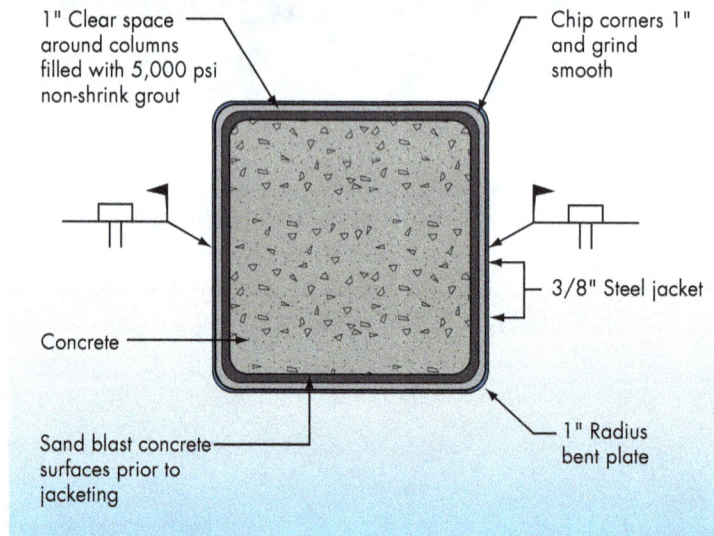

1" Clear space around columns filled with 5,000 psi non-shrink grout

Chip corners 1" and grind smooth

Concrete

3/8" Steel jacket

Sand blast concrete surfaces prior to jacketing

1" Radius bent plate

2.5.3 Checklist for Retrofitting Issues

A Building Vulnerability Assessment Checklist was developed for FEMA 426 and FEMA 452 to help identify structural conditions that may suffer in response to blast loading. Each building in consideration needs to be evaluated by a professional engineer, experienced in the protective design of structures, to determine the ability to withstand blast loading.

In addition, the following questions will help address key retrofitting issues. Issues related to the retrofitting of existing refuge areas (e.g., hallways/corridors, bathrooms, workrooms, laboratory areas, kitchens, and mechanical rooms) that should be considered include the following:

○ **The roof system.** Is the roof system over the proposed refuge area structurally independent of the remainder of the building? If not, is it capable of resisting the expected blast, wind, and debris loads? Are there openings in the roof system for mechanical equipment or lighting that cannot be protected during a blast or high-wind event? It may not be reasonable to retrofit the rest of the proposed shelter area if the roof system is part of a building that was not designed for high-wind load requirements.

○ **The wall system.** Can the wall systems be accessed so that they can be retrofitted for resistance to blast and high-wind pressures and missile impact? It may not be reasonable to retrofit a proposed shelter area to protect openings if the wall systems (load-bearing or non-load-bearing) cannot withstand blast and wind pressures or cannot be retrofitted in a reasonable manner to withstand blast or wind pressures and missile impacts.

○ **Openings.** Are the windows and doors vulnerable to blast and wind pressures and debris impact? Are doors constructed of impact-resistant materials (e.g., steel doors) and secured with six points of connection (typically three hinges and three latching mechanisms)? Are door frames constructed of at least 16-gauge metal and adequately secured to the walls to prevent the complete failure of the door/frame assemblies? Does the building rely on shutter systems for resistance to the effects of hurricanes? There is often only minimal warning time before a CBRE or tornado event; therefore, a shelter design that relies on manually installed shutters is impractical. Automated shutter systems may be considered, but they would require a protected backup power system to ensure that the shutters are closed before an event.

2.6 SHELTERS AND MODEL BUILDING TYPES

This section will provide basic FEMA model building types to describe protective design and structural systems for shelters in the most effective manner. This section is based on FEMA 310, *Handbook for the Seismic Evaluation of Buildings,* which is dedicated to instructing the design professional on how to determine if a building is adequately designed and constructed to resist particular types of forces. Graphics included in this section were prepared for FEMA 454, *Designing for Earthquakes: A Manual for Architects.*

2.6.1   W1, W1a, and W2 Wood Light Frames and Wood Commercial Buildings

Small wood light frame buildings (<3,000 square feet) are single or multiple family dwellings of one or more stories in height (see Figure 2-21). Building loads are light and the framing spans are short. Floor and roof framing consists of closely spaced wood joists or rafters on wood studs. The first floor framing is supported directly on the foundation, or is raised up on cripple studs and post and beam supports. The foundation consists of spread footings constructed of concrete, concrete masonry block, or brick masonry in older construction. Chimneys, when present, consist of solid brick masonry, masonry veneer, or wood frame with internal metal flues. Lateral forces are resisted by wood frame diaphragms and shear walls. Floor and roof diaphragms consist of straight or diagonal wood sheathing, tongue and groove planks, or plywood.

Shearwalls consist of straight or diagonal wood sheathing, plank siding, plywood, stucco, gypsum board, particle board, or fiberboard. Interior partitions are sheathed with plaster or gypsum board.

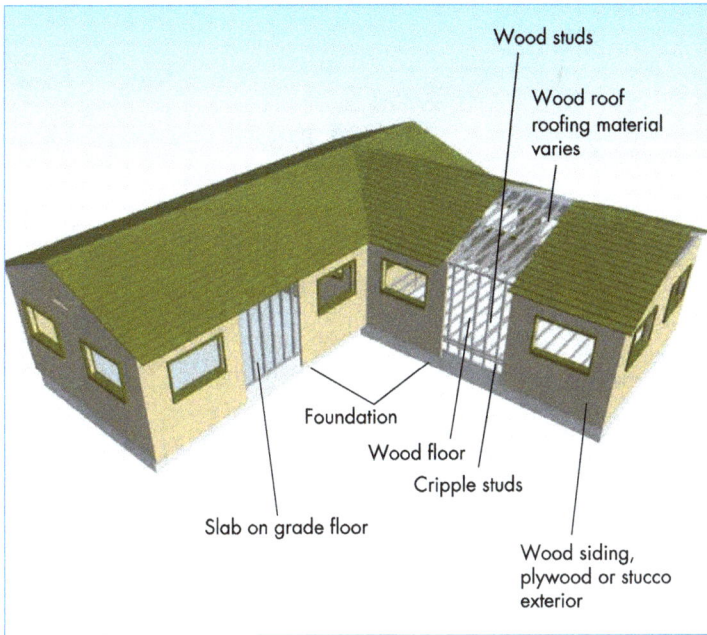

Figure 2-21
W1 wood light frame
< 3,000 square feet

Large wood light frame buildings (> 3,000 square feet) are multi-story, multi-unit residences similar in construction to W1 buildings, but with open front garages at the first story (see Figure 2-22). The first story consists of wood floor framing on wood stud walls and steel pipe columns, or a concrete slab on concrete or concrete masonry block walls.

Wood commercial or industrial buildings with a floor area of 5,000 square feet or more carry heavier loads than light frame construction (see Figure 2-23). In these buildings, the framing spans are long and there are few, if any, interior walls. The floor and roof framing consists of wood or steel trusses, glulam or steel beams, and wood posts or steel columns. Lateral forces are resisted by wood diaphragms and exterior stud walls sheathed with plywood, stucco, plaster, straight or diagonal wood sheathing, or braced with rod bracing. Large openings for storefronts and garages, when present, are framed by post-and-beam framing. Lateral force resistance around openings is provided by steel rigid frames or diagonal bracing.

Figure 2-22
W1a wood light frame
> 3,000 square feet

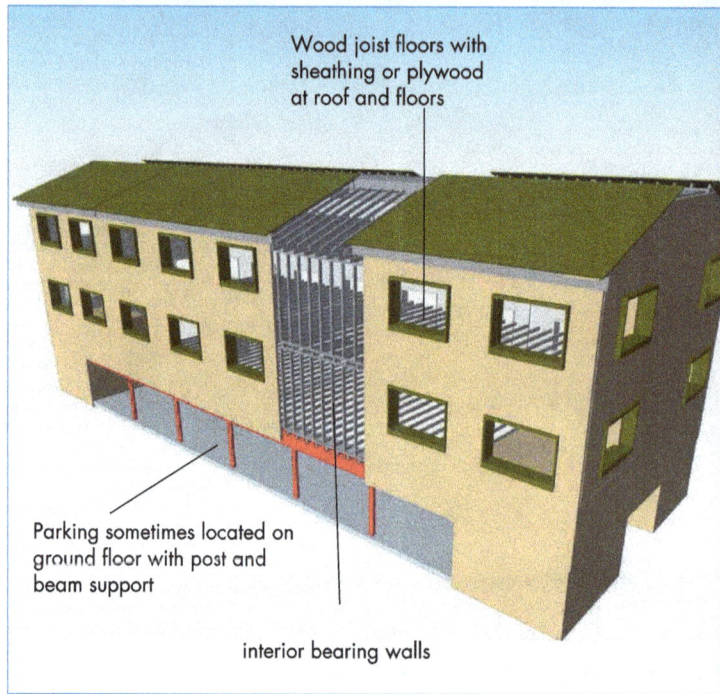

Wood joist floors with sheathing or plywood at roof and floors

Parking sometimes located on ground floor with post and beam support

interior bearing walls

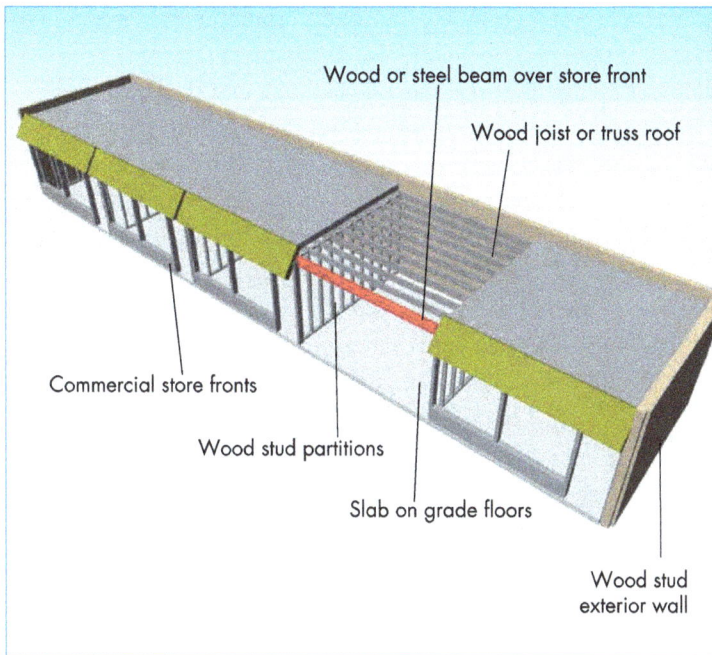

Figure 2-23
W2 wood commercial
buildings

Figure labels: Wood or steel beam over store front; Wood joist or truss roof; Commercial store fronts; Wood stud partitions; Slab on grade floors; Wood stud exterior wall

Light wood frame structures do not possess significant resistance to blast loads although larger wood commercial buildings will be better able to accept these lateral loads than light frame wood construction. These buildings are likely to suffer heavy damage in response to 50 pounds of TNT at a stand-off distance of 20 to 50 feet. A shelter would best be located in a basement where the protection to blast loading would be provided by the surrounding soil. Large explosive detonations in close proximity to the building will not only destroy the superstructure, but the effects of ground shock are likely to fail the foundation walls as well; therefore, protected spaces should be located interior to the building. Locating the shelter on the ground floor, for slab on grade structures, provides the maximum number of floors between occupants and possible roof debris. Debris catch systems may be installed beneath roof rafters of single-story buildings; however, the effectiveness of the debris catch system will be limited if the zone of roof damage is extensive.

Metal stud blast walls built within the existing building may be used to supplement the enclosure; however, in order for these walls to develop their resistance to lateral loads, they must be anchored to an existing structure. Windows enclosing the selected shelter must either be laminated or treated with an anti-shatter film. Either the laminated glass or the anti-shatter film should be anchored to the surrounding wall with a system that can develop but not overwhelm the capacity of the wall. A conservative estimate of the ultimate capacity of an existing wall may be determined, in the absence of actual design information, by scaling the code specified wind pressures with the appropriate factor of safety.

2.6.2 S1, S2, and S3 Steel Moment Frames, Steel Braced Frames, and Steel Light Frames

Steel moment frame and braced frame buildings with cast-in-place concrete slabs or metal deck with concrete fill supported on steel beams, open web joists, or steel trusses are well suited for a hardened shelter construction. Lateral forces in steel moment frame buildings are resisted by means of rigid or semi-rigid beam-column connections (see Figure 2-24). When all connections are moment-resisting connections, the entire frame participates in lateral force resistance. When only selected connections are moment-resisting connections, resistance is provided along discrete frame lines. Columns are oriented so that each principal direction of the building has columns resisting forces in strong axis bending. Diaphragms consist of concrete or metal deck with concrete fill and are stiff relative to the frames. Walls may consist of metal panel curtainwalls, glazing, brick masonry, or precast concrete panels. When the interior of the structure is finished, frames are concealed by ceilings, partition walls, and architectural column furring. Foundations consist of concrete spread footings or deep pile foundations.

Lateral forces in steel braced frame buildings are resisted by tension and compression forces in diagonal steel members (see Figure 2-25). When diagonal brace connections are concentric to

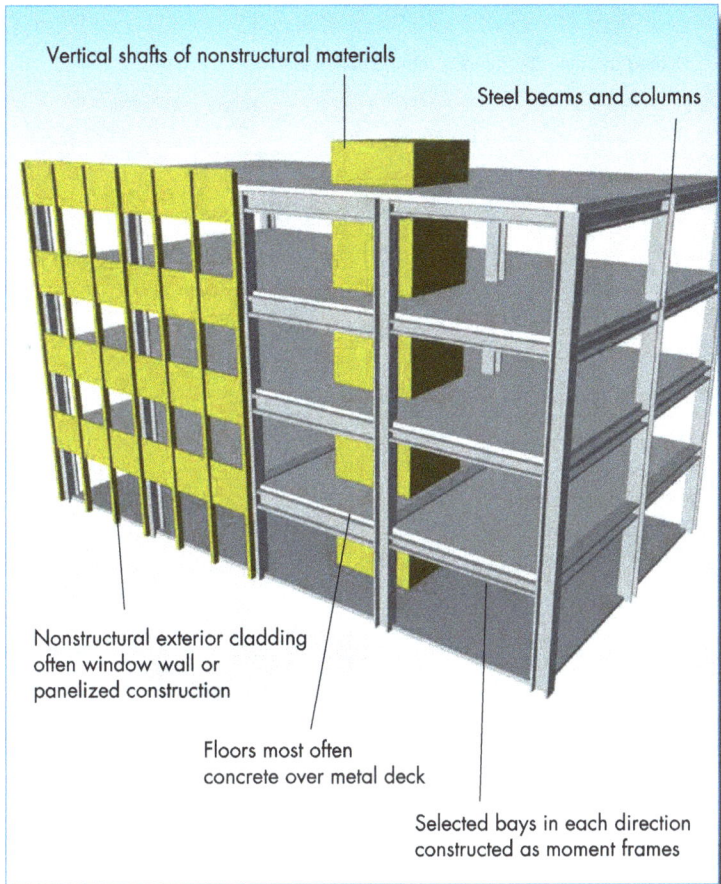

Figure 2-24
S1 steel moment frames

Vertical shafts of nonstructural materials

Steel beams and columns

Nonstructural exterior cladding
often window wall or
panelized construction

Floors most often
concrete over metal deck

Selected bays in each direction
constructed as moment frames

beam column joints, all member stresses are primarily axial. When diagonal brace connections are eccentric to the joints, members are subjected to bending and axial stresses. Diaphragms consist of concrete or metal deck with concrete fill and are stiff relative to the frames. Walls may consist of metal panel curtainwalls, glazing, brick masonry, or precast concrete panels. When the interior of the structure is finished, frames are concealed by ceilings, partition walls, and architectural furring. Foundations consist of concrete spread footings or deep pile foundations.

Figure 2-25
S2 steel braced frames

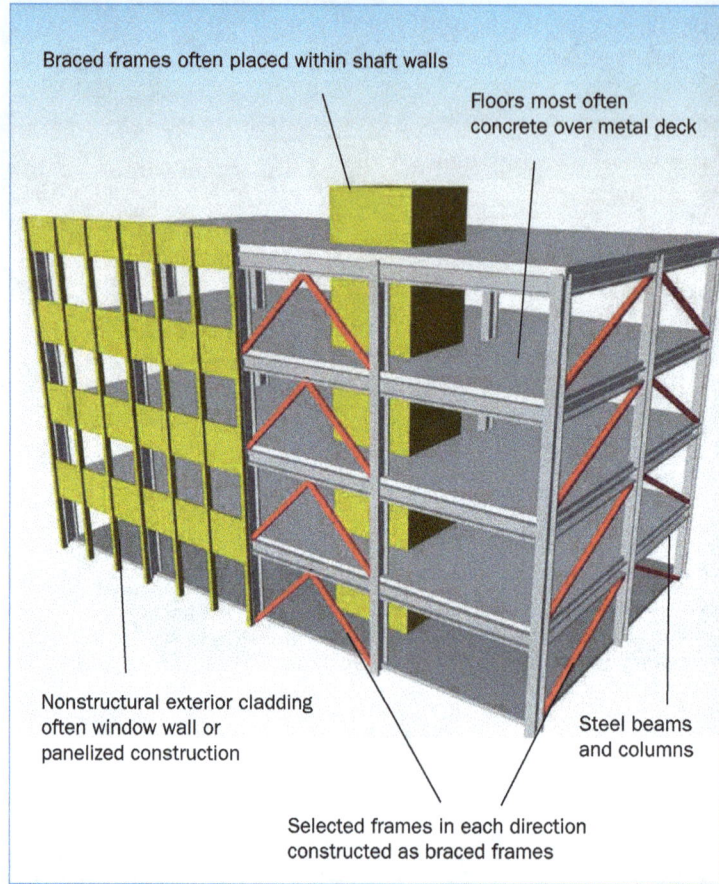

Braced frames often placed within shaft walls

Floors most often concrete over metal deck

Nonstructural exterior cladding often window wall or panelized construction

Steel beams and columns

Selected frames in each direction constructed as braced frames

Light frame steel structures are pre-engineered and prefabricated with transverse rigid steel frames (see Figure 2-26). They are one-story in height and the roof and walls consist of lightweight metal, fiberglass, or cementitious panels. The frames are designed for maximum efficiency and the beams and columns consist of tapered, built-up sections with thin plates. The frames are built in segments and assembled in the field with bolted or welded joints. Lateral forces in the transverse direction are resisted by the rigid frames. Lateral forces in the longitudinal direction are resisted by wall panel shear elements or rod bracing. Diaphragm forces are resisted by untopped metal deck, roof panel shear elements, or a system of tension-only rod bracing.

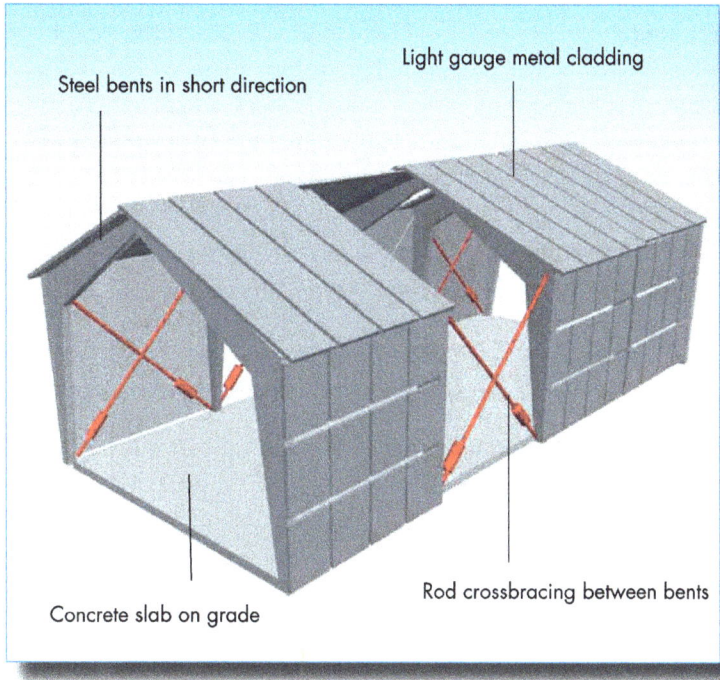

Figure 2-26
S3 steel light frames

Steel bents in short direction

Light gauge metal cladding

Concrete slab on grade

Rod crossbracing between bents

Steel moment frame structures provide excellent ductility and redundancy in response to blast loading. Steel braced frames may similarly be designed to resist high intensity blast loads; however, they are less effective in resisting the progression of collapse following the loss of a primary load-bearing element. As a result, first floor steel columns of existing buildings may be concrete encased and first floor splices may be reinforced to increase their resistance to local failure that could precipitate a progression of collapse. The exterior façade represents the most fragile element and is likely to be severely damaged in response to an exterior detonation. Debris may be minimized by means of reinforced masonry, sufficiently detailed precast panels, or laminated glass façade. Nevertheless, a shelter within steel frame buildings would best be located within interior space or a building core. Hardened interior partitions may easily be constructed and anchored to existing floor slabs, and lightweight metal gauge walls may be used to retrofit existing buildings. Metal deck roofs with rigid insulation supported by bar joist structural elements possess minimal resistance to blast pressures. The additional mass, stiffness, and strength of metal deck roofs with concrete fill make them much better able to resist the effects of direct blast loading and the subsequent rebound. Therefore, lightweight roofs of light frame steel structures are likely to be severely damaged in response to any sizable blast loading and a shelter should either be located in the basement or as interior to the building (as far from the exterior façade) as possible.

2.6.3   S4 and S5 Steel Frames with Concrete Shearwalls and Infill Masonry Walls

Steel frame buildings with concrete or infill masonry shear walls with cast-in-place concrete slabs or metal deck with concrete fill supported on steel beams, open web joists, or steel trusses are well suited for a hardened shelter construction. When lateral forces are resisted by cast-in-place concrete shear walls, the walls carry their own weight. In older construction, the steel frame is designed for vertical loads only. In modern dual systems, the steel moment frames are designed to work together with the concrete shear walls in proportion to their relative rigidity (see Figure 2-27). In the case of a dual system, the walls should be evaluated under this building type and the frames should be evaluated under S1 steel moment frames. Diaphragms consist of concrete or metal deck with or without concrete fill. The steel frame may provide a secondary lateral-force-resisting system, depending on the stiffness of the frame and the moment capacity of the beam-column connections.

STRUCTURAL DESIGN CRITERIA

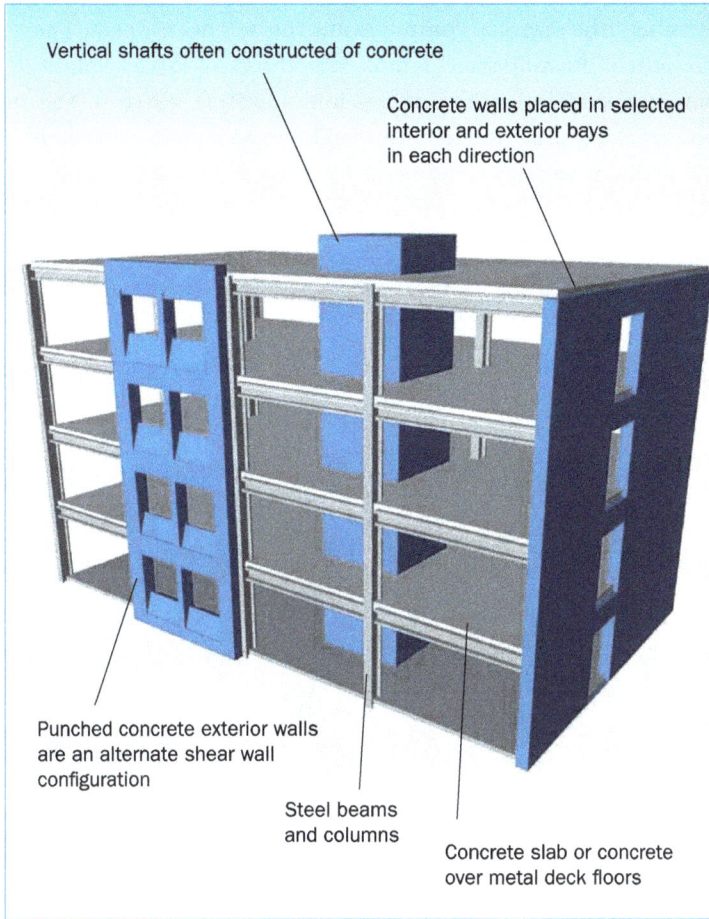

Vertical shafts often constructed of concrete

Concrete walls placed in selected interior and exterior bays in each direction

Punched concrete exterior walls are an alternate shear wall configuration

Steel beams and columns

Concrete slab or concrete over metal deck floors

Figure 2-27
S4 steel frames with concrete shearwalls

Steel frames with infill masonry walls is an older type of building construction (see Figure 2-28). The walls consist of infill panels constructed of solid clay brick, concrete block, or hollow clay tile masonry. Infill walls may completely encase the frame members, and present a smooth masonry exterior with no indication of the frame. The lateral resistance of this type of construction depends on the interaction between the frame and infill panels. The combined behavior is more like a shear wall structure than a frame structure. Solidly infilled masonry panels form diagonal compression struts between the intersections of the frame members. If the walls are offset from the frame and do not fully engage the frame members, the diagonal compression struts will not develop. The strength of the infill panel is limited by the shear capacity of the masonry bed joint or the compression capacity of the strut. The post-cracking strength is determined by an analysis of a moment frame that is partially restrained by the cracked infill. The diaphragms consist of concrete floors and are stiff relative to the walls.

Figure 2-28
S5 steel frames with infill masonry walls

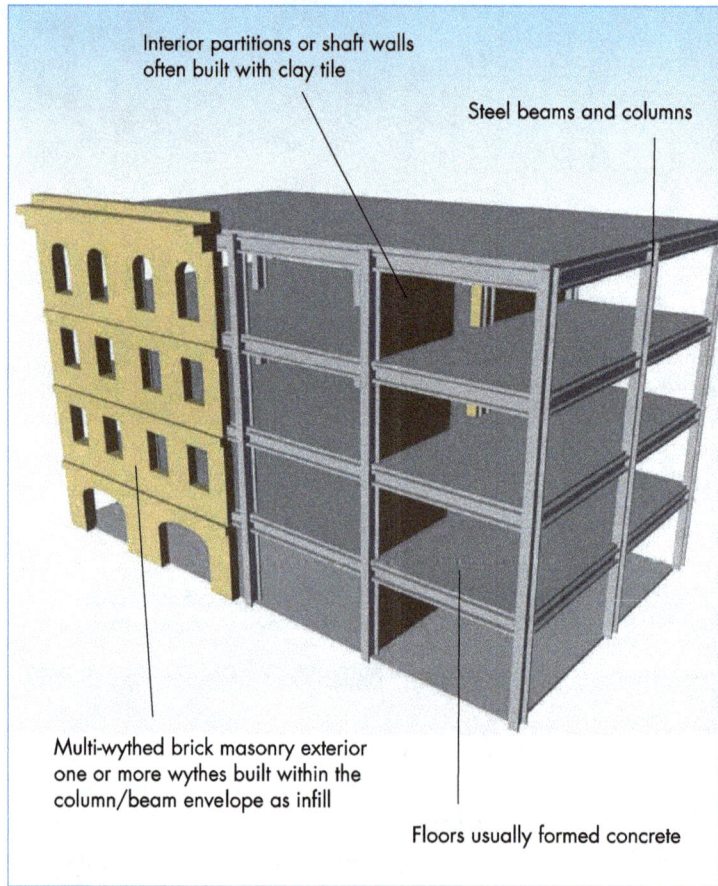

Interior partitions or shaft walls often built with clay tile

Steel beams and columns

Multi-wythed brick masonry exterior one or more wythes built within the column/beam envelope as infill

Floors usually formed concrete

Steel frame structures with either concrete shear walls or infill masonry walls are not moment connected; therefore, the frame is more vulnerable to collapse resulting from the loss of a column. As a point of reference, steel moment frame buildings with lightly reinforced CMU infill walls are likely to suffer heavy damage in response to 500 pounds of TNT at a stand-off distance of 50 feet or less. The first floor steel columns of existing buildings may be concrete encased and first floor splices may be reinforced to increase their resistance to local failure that could precipitate a progression of collapse. The exterior façade is likely to be damaged in response to an exterior detonation and debris may be minimized by means of reinforced masonry, sufficiently detailed precast panels, or laminated glass façade. Nevertheless, a shelter within these buildings would best be located within interior space or a building core, preferably enclosed on one or more sides by the shear walls. Existing masonry infill walls may be retrofitted to supplement existing reinforcement by either grouting cables within holes cored within the walls or with a spray-on application of a shotcrete and welded wire fabric or a polyurea debris catch membrane. Alternatively, hardened interior partitions may easily be constructed and anchored to existing floor slabs, and lightweight metal stud walls may be used to retrofit existing buildings.

2.6.4 C1, C2, and C3 Concrete Moment Frames, Concrete and Infill Masonry Shearwalls – Type 1 Bearing Walls and Type 2 Gravity Frames

These buildings consist of a frame assembly of cast-in-place concrete beams and columns. Floor and roof framing consists of cast-in-place concrete slabs, concrete beams, one-way joists, two-way waffle joists, or flat slabs. Lateral forces are resisted by concrete moment frames that develop their stiffness through monolithic beam-column connections (see Figure 2-29). In older construction, or in regions of low seismicity, the moment frames may consist of the column strips of two-way flat slab systems. Modern frames in regions of high seismicity have joint reinforcing, closely spaced ties, and special detailing to provide ductile performance. This detailing is not present in older construction. Foundations consist of concrete spread footings or deep pile foundations.

Figure 2-29
C1 concrete moment
frames

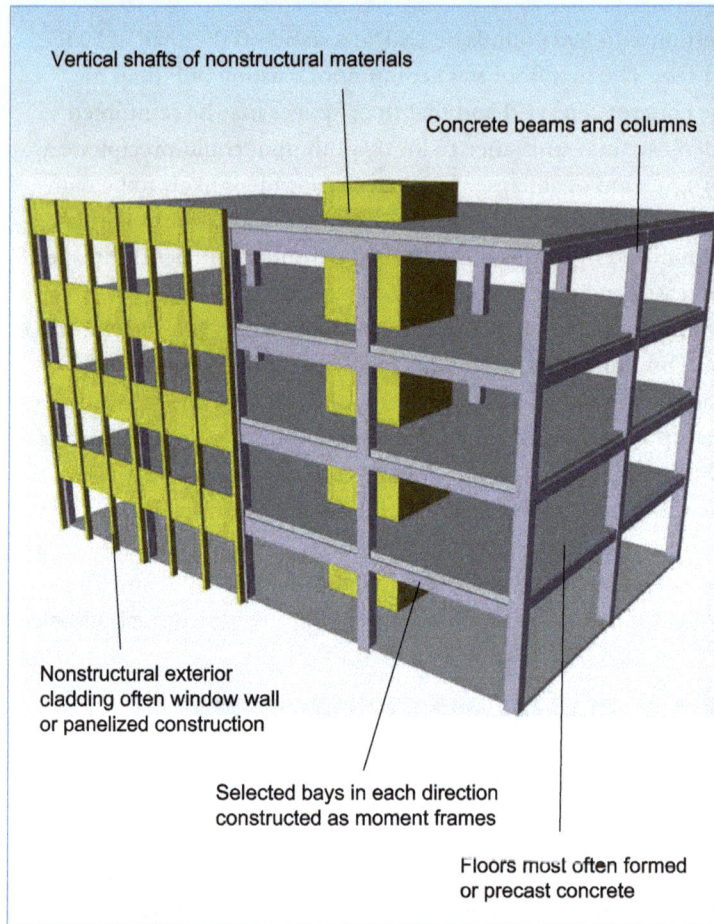

Vertical shafts of nonstructural materials

Concrete beams and columns

Nonstructural exterior
cladding often window wall
or panelized construction

Selected bays in each direction
constructed as moment frames

Floors most often formed
or precast concrete

Concrete and infill masonry shearwall building systems have
floor and roof framing that consists of cast-in-place concrete
slabs, concrete beams, one-way joists, two-way waffle joists, or flat
slabs. Floors are supported on concrete columns or bearing walls.
Lateral forces are resisted by cast-in-place concrete shear walls
or infill panels constructed of solid clay brick, concrete block,
or hollow clay tile masonry (see Figures 2-30, 2-31, and 2-32). In
older construction, cast-in-place shear walls are lightly reinforced,
but often extend throughout the building. In more recent construction, shear walls occur in isolated locations and are more
heavily reinforced with boundary elements and closely spaced ties

to provide ductile performance. The diaphragms consist of concrete slabs and are stiff relative to the walls. Foundations consist of concrete spread footings or deep pile foundations. The seismic performance of infill panel construction depends on the interaction between the frame and infill panels. The combined behavior is more like a shear wall structure than a frame structure. If the infilled masonry panels are in line with the frame, they form diagonal compression struts between the intersections of the frame members; otherwise, the diagonal compression struts will not develop. The strength of the infill panel is limited by the shear capacity of the masonry bed joint or the compression capacity of the strut. The post-cracking strength is determined by an analysis of a moment frame that is partially restrained by the cracked infill. The shear strength of the concrete columns, after cracking of the infill, may limit the semiductile behavior of the system.

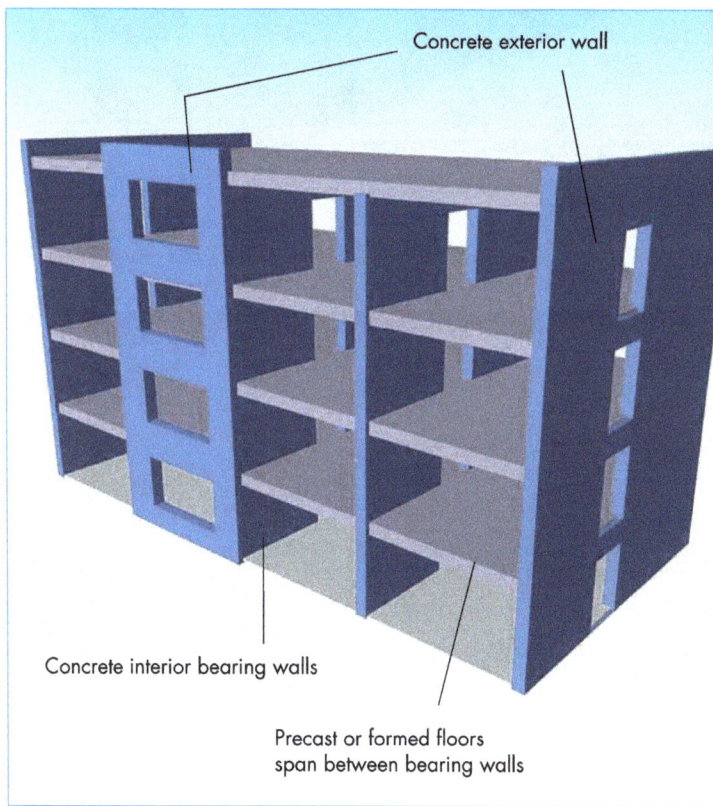

Concrete exterior wall

Concrete interior bearing walls

Precast or formed floors
span between bearing walls

Figure 2-30
C2 concrete shearwalls
– type 1 bearing walls

Figure 2-31
C2 concrete shearwalls
– type 2 gravity frames

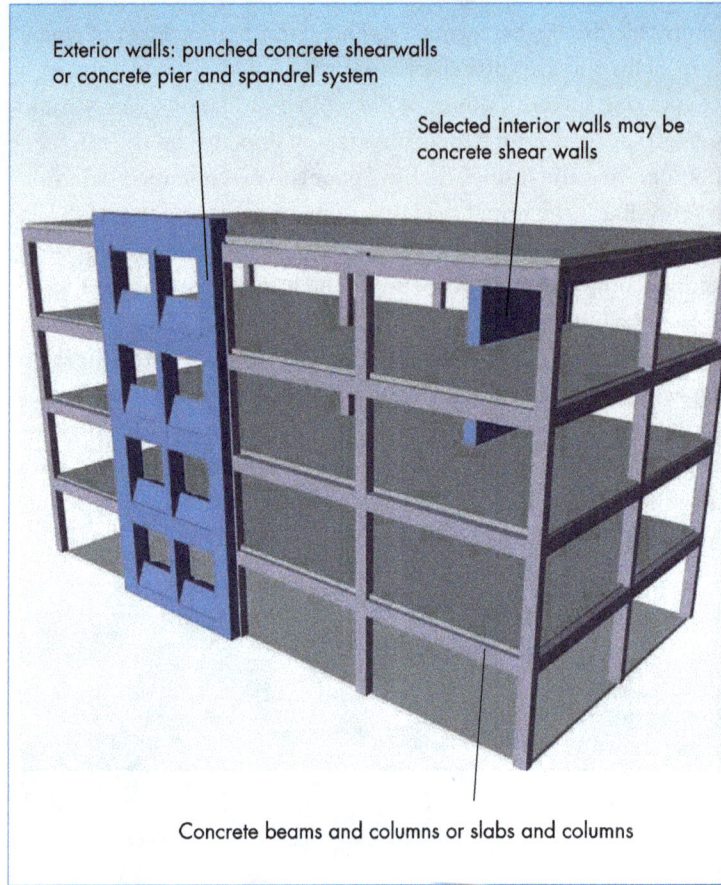

Exterior walls: punched concrete shearwalls
or concrete pier and spandrel system

Selected interior walls may be
concrete shear walls

Concrete beams and columns or slabs and columns

STRUCTURAL DESIGN CRITERIA

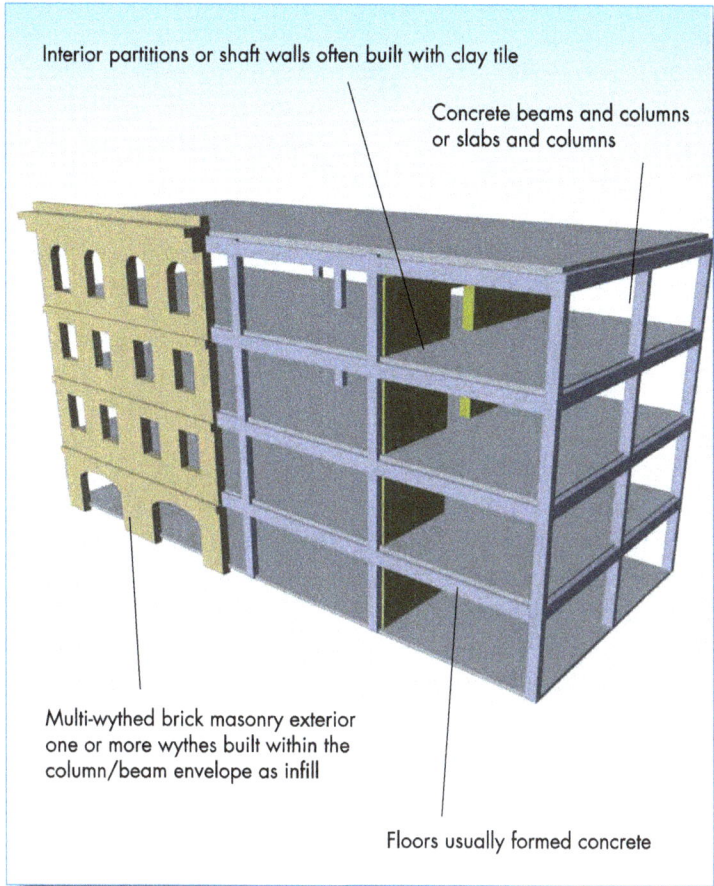

Interior partitions or shaft walls often built with clay tile

Concrete beams and columns or slabs and columns

Figure 2-32
C3 concrete frames with infill masonry shearwalls

Multi-wythed brick masonry exterior one or more wythes built within the column/beam envelope as infill

Floors usually formed concrete

Unless sited in a seismic zone, concrete frame structures are not typically designed and detailed to develop large inelastic deformations and withstand significant load reversals. As a point of reference, a building with 8-inch thick reinforced concrete load-bearing exterior walls and interior columns is likely to suffer heavy damage in response to 500 pounds of TNT at a distance of 70 feet or less. The exterior façade is likely to be damaged in response to an exterior detonation and debris may be minimized by means of reinforced masonry, sufficiently detailed precast panels, or laminated glass façade. Nevertheless, a shelter within concrete frame and shearwall buildings would best be located within interior space or a building core, preferably enclosed on one or more sides by the shear walls. Existing masonry infill walls may be retrofitted to supplement existing reinforcement by either grouting cables within holes cored within the walls or with a spray-on application of a shotcrete and welded wire fabric or a polyurea debris catch membrane. Alternatively, hardened interior partitions may easily be constructed and anchored to existing floor slabs, and lightweight metal stud walls may be used to retrofit existing buildings.

2.6.5   PC1 and PC2 Tilt-up Concrete Shearwalls and Precast Concrete Frames and Shearwalls

Tilt-up concrete buildings are one or more stories in height and have precast concrete perimeter wall panels that are cast on site and tilted into place (see Figure 2-33). Floor and roof framing consists of wood joists, glulam beams, steel beams, open web joists, or precast plank sections. Framing is supported on interior steel or concrete columns and perimeter concrete bearing walls. The floors consist of wood sheathing, concrete over form deck, or composite concrete slabs. Roofs are typically untopped metal deck, but may contain lightweight concrete fill. Lateral forces are resisted by the precast concrete perimeter wall panels. Wall panels may be solid, or have large window and door openings that cause the panels to behave more as frames than as shear walls. In older construction, wood framing is attached to the walls with wood ledgers. Foundations typically consist of concrete spread footings or deep pile foundations.

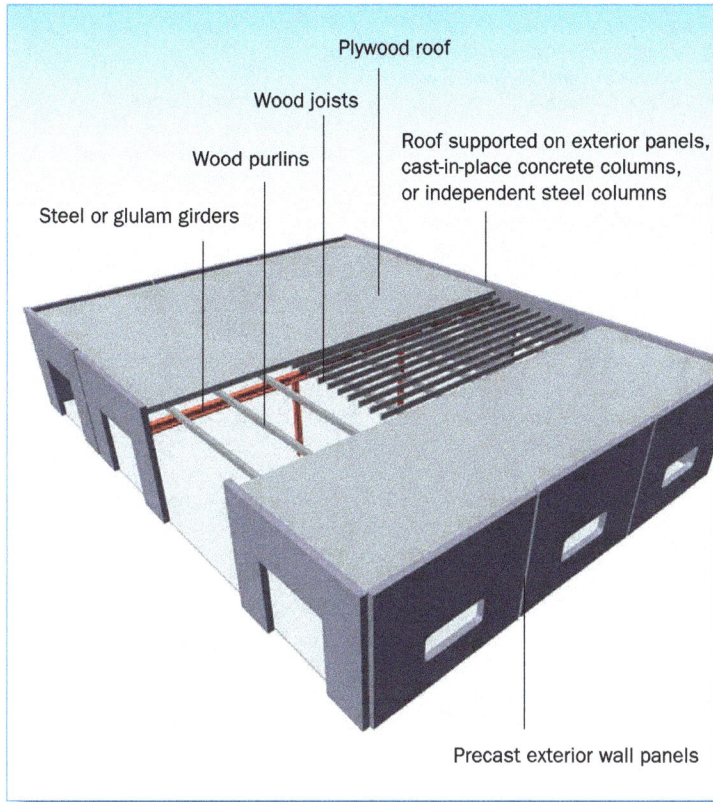

Figure 2-33
PC1 tilt-up concrete
shearwalls

Labels in figure:
- Plywood roof
- Wood joists
- Wood purlins
- Steel or glulam girders
- Roof supported on exterior panels, cast-in-place concrete columns, or independent steel columns
- Precast exterior wall panels

Precast concrete frames and shearwalls consist of precast concrete planks, tees, or double-tees supported on precast concrete girders and precast columns (see Figure 2-34). Lateral forces are resisted by precast or cast-in-place concrete shear walls. Diaphragms consist of precast elements interconnected with welded inserts, cast-in-place closure strips, or reinforced concrete topping slabs.

Figure 2-34
PC2 precast concrete
frames and shearwalls

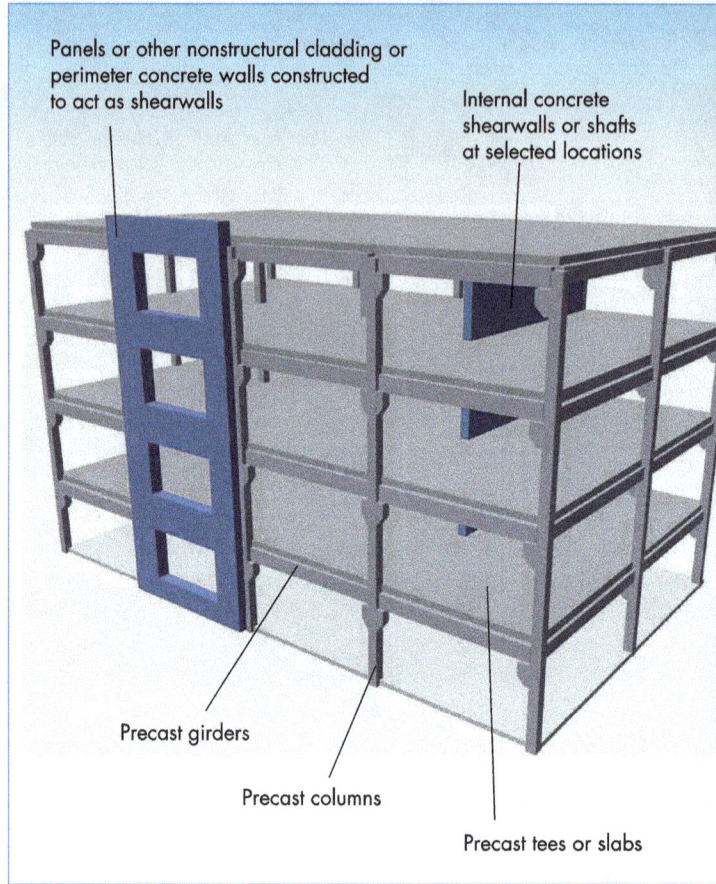

Panels or other nonstructural cladding or
perimeter concrete walls constructed
to act as shearwalls

Internal concrete
shearwalls or shafts
at selected locations

Precast girders

Precast columns

Precast tees or slabs

Precast construction benefits from higher quality wall and frame components than cast-in-place structures; however, it lacks the continuity of construction present in these systems. The resistance blast loading depends, to a great extent, on the mechanical connections between the components. Designers must consider the blast loading effects when designing and detailing these connections. A shelter would best be located in a basement where the protection to blast loading would be provided by the surrounding soil. Large explosive detonations in close proximity to the building will not only destroy the superstructure, but the effects of ground shock are likely to fail the foundation walls as well; therefore, protected spaces should be located interior to the

STRUCTURAL DESIGN CRITERIA

building. Locating the shelter in the basement, for slab on grade buildings, provides the maximum number of floors between occupants and possible roof debris. Debris catch systems may be installed beneath roof rafters of single-story buildings; however, the effectiveness of the debris catch system will be limited if the zone of roof damage is extensive.

Metal stud blast walls built within the existing building may be used to supplement the enclosure; however, in order for these walls to develop their resistance to lateral loads, they must be anchored to an existing structure. Windows enclosing the selected shelter must either be laminated or treated with an anti-shatter film. Either the laminated glass or the anti-shatter film should be anchored to the surrounding wall with a system that can develop, but not overwhelm the capacity of the wall. A conservative estimate of the ultimate capacity of an existing wall may be determined, in the absence of actual design information, by scaling the code specified wind pressures with the appropriate factor of safety.

2.6.6   RM1 and RM2 Reinforced Masonry Walls with Flexible Diaphragms or Stiff Diaphragms and Unreinforced Masonry (URM) Load-bearing Walls

These buildings have bearing walls that consist of reinforced brick or concrete block masonry. Wood floor and roof framing consists of wood joists, glulam beams, and wood posts or small steel columns. Steel floor and roof framing consists of steel beams or open web joists, steel girders, and steel columns. Lateral forces are resisted by the reinforced brick or concrete block masonry shear walls. Diaphragms consist of straight or diagonal wood sheathing, plywood, or untopped metal deck, and are flexible relative to the walls (see Figure 2-35). Foundations consist of brick or concrete spread footings.

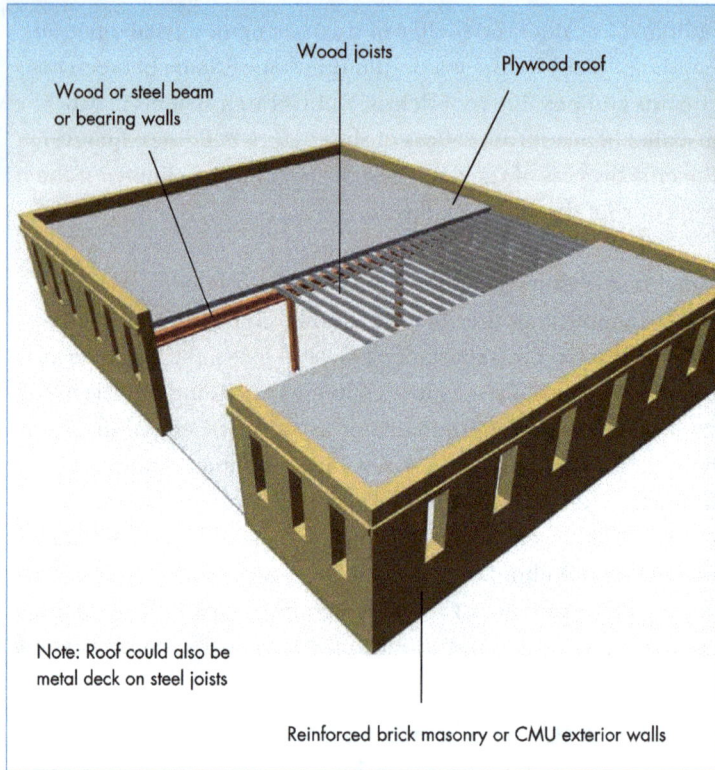

Figure 2-35
RM1 reinforced masonry
walls with flexible
diaphragms

Wood or steel beam
or bearing walls

Wood joists

Plywood roof

Note: Roof could also be
metal deck on steel joists

Reinforced brick masonry or CMU exterior walls

Buildings with reinforced masonry walls and stiff diaphragms are
similar to RM1 buildings, except the diaphragms consist of metal
deck with concrete fill, precast concrete planks, tees, or double-
tees, with or without a cast-in-place concrete topping slab, and
are stiff relative to the walls (see Figure 2-36). The floor and roof
framing is supported on interior steel or concrete frames or interior reinforced masonry walls.

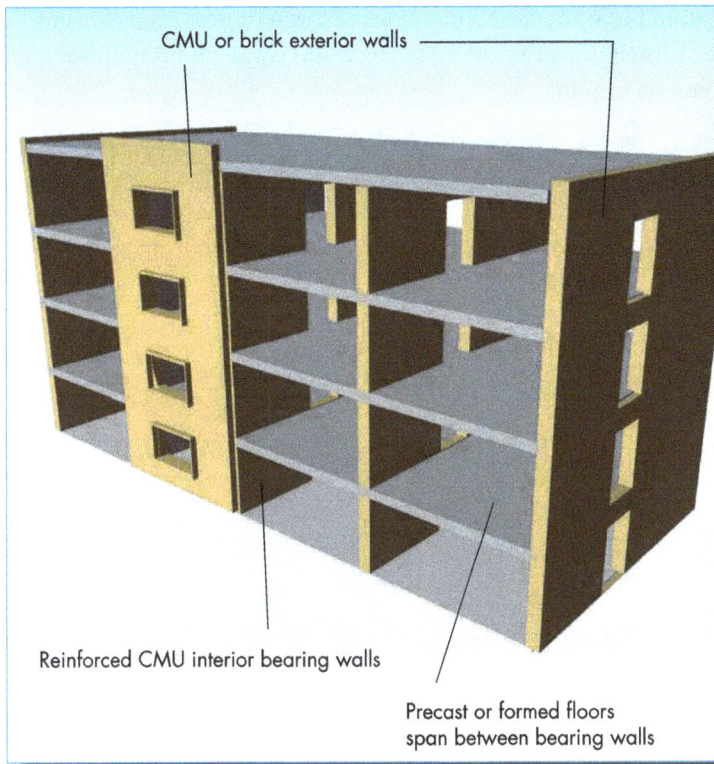

CMU or brick exterior walls

Reinforced CMU interior bearing walls

Precast or formed floors
span between bearing walls

Figure 2-36
RM2 reinforced masonry
walls with stiff diaphragms

Unreinforced load-bearing masonry buildings often contain perimeter bearing walls and interior bearing walls made of clay brick masonry (see Figure 2-37). In older construction, floor and roof framing consists of straight or diagonal lumber sheathing supported by wood joists, on posts and timbers. In more recent construction, floors consist of structural panel or plywood sheathing rather than lumber sheathing. The diaphragms are flexible relative to the walls. When they exist, ties between the walls and diaphragms consist of bent steel plates or government anchors embedded in the mortar joints and attached to framing. Foundations consist of brick or concrete spread footings. As a variation, some URM buildings have stiff diaphragms relative to the unreinforced masonry walls and interior framing. In older construction or large, multi-story buildings, diaphragms may consist of cast-in-place concrete. In regions of low seismicity, more recent construction consists of metal deck and concrete fill supported on steel framing.

Figure 2-37
URM load-bearing walls

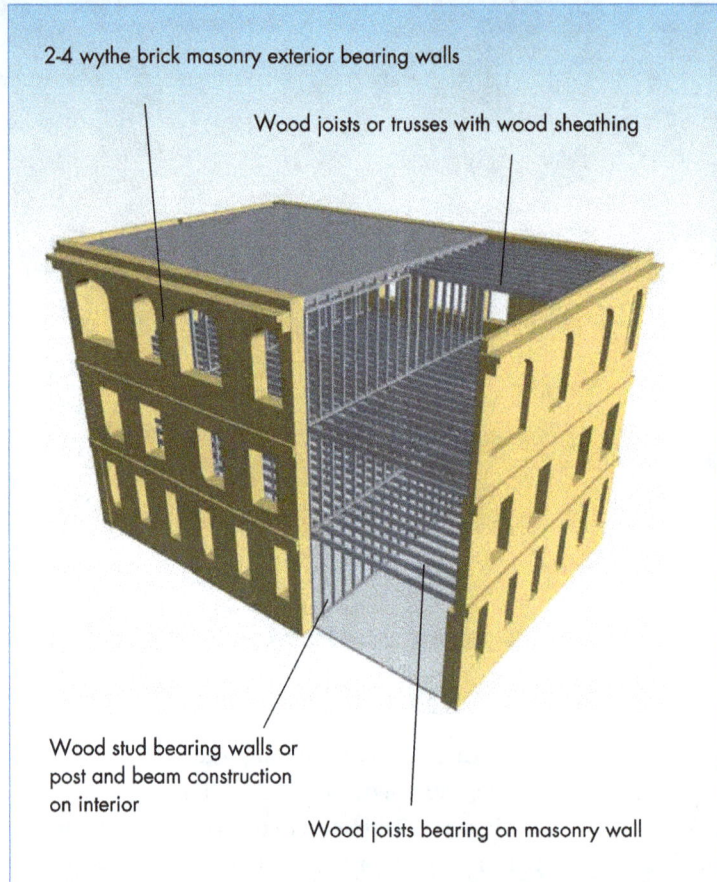

Unless sited in a seismic zone, reinforced masonry structures are not typically detailed to develop significant inelastic deformations and withstand significant load reversals. Unreinforced masonry structures are extremely brittle. As a point of reference, a reinforced masonry building with 8-inch thick reinforced CMU exterior walls is likely to suffer heavy damage in response to 500 pounds of TNT at a distance of 150 feet or less. An unreinforced masonry building with reinforced CMU pilasters will suffer heavy damage in response to 500 pounds of TNT at a distance of 250 feet or less. At these loads, the structure supported by the load-bearing masonry wall is likely to suffer localized collapse. Grout and additional reinforcement may be inserted within the cores of existing masonry walls; however, a stiffened steel panel provides the most effective way to restrain the debris and assume the gravity loads following the loss of load carrying capacity within the wall. A shelter within these buildings would best be located within interior space or a building core, preferably enclosed on one or more sides by the shear walls.

2.6.7   Conclusions

Despite the various types of construction, the following protective measures may be used to establish a hardened space that will limit the extent of debris resulting from an explosive event. A shelter is best located within interior space or a building core at the lowest levels of a building or on the ground floor for a slab on grade structure. A debris catch system should be installed beneath the roof rafters of a single-story building. The exterior façade should be either reinforced masonry or precast panels and windows should either be laminated or treated with an anti-shatter film that is anchored to the surrounding walls. First floor steel columns may be concrete encased and first floor splices may be reinforced. Existing masonry infill walls may be retrofitted by either grouting cables within holes cored within the walls or with a spray-on application of a shotcrete and welded wire fabric or a polyurea debris catch membrane. Hardened interior partitions, such as metal stud blast walls, may be used to enclose the shelter and these walls should be anchored to an existing structure. A stiffened steel panel may be constructed interior to existing load-bearing masonry walls.

2.7   CASE STUDY: BLAST-RESISTANT SAFE ROOM

Consider the example of a safe room established in the stairwell of a multi-story office building: it may be assumed the original construction did not provide for reinforced masonry or reinforced concrete enclosures. To achieve the greatest stand-off distance and isolate the safe room from a vehicle-borne explosive threat, the stairwell should be interior to the structure. This will provide the maximum level of protection from an undefined explosive threat. Although it is common to place emergency stairs within the building core, one can only reasonably expect a reinforced concrete or reinforced masonry stair enclosure for a shearwall lateral resisting structural system. Due to the large difference in weight and constructability, a stud wall with gypsum board stair enclosure will be routinely used in lieu of reinforced masonry or concrete for framed construction. The stair enclosures may therefore be designed or upgraded to include 16-gauge sheet metal supported by 18-gauge steel studs that are attached web to web (back to back). These walls must be adequately anchored to the existing floor slabs to develop the plastic capacity of the studs acting both in flexure and in tension. Alternatively, fully grouted reinforced masonry stairwell enclosures, #4 bars in each cell, may be specified. The masonry walls must be adequately anchored to the existing floor slabs to develop the ultimate lateral resistance of the wall in order to transfer the reaction loads into the lateral resisting system of the building. Doors to the stairway enclosures are to be hollow steel or steel clad, such as 14-gauge steel doors with 20-gauge ribs, with pressed steel frames; double doors should utilize a center stile. Doors should open away from the safe room and be securely anchored to the wall construction, locally reinforced around the door.

Any windows within the stairwell enclosures are to contain laminated glass, utilizing 0.060 PVB, that is adhered within the mullions with a ½-inch bead of structural silicone. The mullions are to be anchored into the surrounding walls to develop the full capacity of the glazing materials. Alternatively, a 7-mil anti-shatter film may be applied to existing windows and mechanically attached to the surrounding mullions to develop the full capacity of the film. A wet glazed attachment of the film may alternatively be applied; however, this provides a less reliable bond to the existing mullions.

Floor slabs within an interior stairwell will be isolated from the most direct effects of an exterior explosive event and will not be subjected to significant uplift pressures resulting from an exterior explosive event. Nevertheless, for new construction, floor slabs should be designed to withstand a net upward load of magnitude equal to the dead load plus half the live load for the floor system.

For new construction, the structural frames are to be sufficiently tied as to provide alternate load paths to surrounding columns or beams in the event of localized damage. These tie forces should, at a minimum, conform to the DoD Unified Facilities Criteria (UFC) 4-023-03, *Design of Buildings to Prevent Progressive Collapse.* For reinforced concrete structures, seismic hooks and seismic development lengths, as specified in Chapter 21 of the American Concrete Institute (ACI) 318-05, should be used to anchor and develop steel reinforcement. Internal tie reinforcement should be distributed in two perpendicular directions and be continuous from one edge of the floor or roof to the far edge of the floor or roof, using lap splices, welds, or mechanical splices. In order to redistribute the forces that may develop, the internal ties must be anchored to the peripheral ties at each end (see Figure 2-38). Steel structures must be similarly tied, and each column must be effectively held in position by means of horizontal ties in two orthogonal directions at each principal floor level supported by that column.

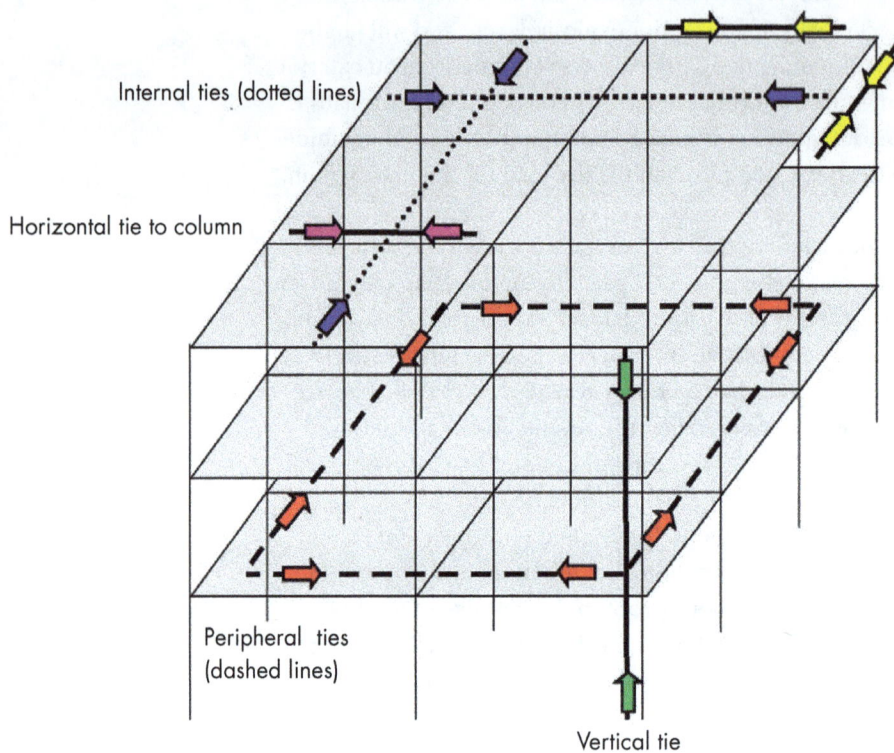

Internal ties (dotted lines)

Horizontal tie to column

Peripheral ties
(dashed lines)

Vertical tie

Figure 2-38   Schematic of tie forces in a frame structure

3.1 OVERVIEW

This chapter describes how to add CBR protection capability to a shelter or safe room.

A CBR safe room protects its occupants from contaminated air outside it by providing clean, breathable air in two ways: (1) by trapping air inside the room and minimizing the air exchange (an unventilated safe room) and (2) by passing contaminated air through a filter to purify it as it is supplied to the room (a ventilated safe room).

Unventilated safe rooms that are tightly sealed cannot be occupied for long periods without the risk of high carbon dioxide levels. This constraint does not apply to ventilated safe rooms, which can be designed to provide filtered and conditioned fresh air at any desired rate. Ventilated safe rooms can therefore be used on a routine basis, although most are designed as standby systems, not for continuous, routine use.

Obtaining protection from an unventilated safe room can be as simple as selecting a relatively tight room, entering it, and closing the door. This procedure is commonly referred to as expedient sheltering-in-place. In this simple form, a safe room protects its occupants by retaining a volume of clean air and minimizing the infiltration of contaminated outdoor air. In practice, however, a safe room is not perfectly tight. The natural forces of wind and buoyancy act on small, distributed leakage paths to exchange air between the inside and outside.

As contaminated air infiltrates a safe room, the level of protection to the occupants diminishes with time. With infiltration in a sustained exposure, the concentration of toxic vapor, gas, or aerosol in the safe room may actually exceed the concentration outdoors because the sealed safe room tends to retain the airborne contaminants when they infiltrate. Once contaminants have entered, they are released slowly after the outdoor hazard has passed. To minimize the hazard of this retention, an unventilated safe room requires two actions to achieve protection:

○ The first is to tighten the safe room, to reduce the indoor-outdoor air exchange rate, before the hazardous plume arrives. This is done by closing doors and windows and turning off fans, air conditioners, and combustion heaters.

○ The second is to aerate, to increase the indoor-outdoor air exchange rate as soon as the plume has passed. This is done by opening doors and windows and turning on all fans and/or exiting the building into clean outdoor air.

The protection a safe room provides can be increased substantially by adding high-efficiency air filtration. Filtration is employed in two different ways to remove contaminants from the air as it enters the safe room or to remove contaminants as air is circulated within the room. The two ways of incorporating filtration, or not incorporating it at all, yield three general configurations or classes of safe rooms, designated Classes 1, 2, and 3. Table 3-1 shows these three classes and summarizes their advantages and limitations. A common element of all three is a tight enclosure. The three classes differ in whether/how air filtration is applied, resulting in differences in cost, level of protection, and duration of protection.

> When planning to use sheltering as a protective action from a CBR release, it is important to consider when the sheltering process should end. Sheltering provides protection by reducing the airflow from the outside that could contain potentially contaminated air. However, some leakage into the shelter may occur and air inside the shelter may reach unacceptable levels of contamination. It is important to leave the shelter when the threat outside has passed.

○ **Class 1**. In a Class 1 Safe Room, air is drawn from outside the room, filtered, and discharged inside the room at a rate sufficient to produce an internal pressure. The safe room is thus ventilated with filtered air, eliminating the constraints related to carbon dioxide accumulation. The internal pressure produced with filtered air prevents infiltration of outside air through leakage paths.

○ **Class 2**. This class also includes air filtration, but with little or no internal pressure. Without positive pressure, the safe room does not prevent the infiltration of contaminated air. A Class 2 Safe Room may be ventilated or unventilated. In an unventilated Class 2 Safe Room, air is drawn from inside the safe room, filtered, and discharged inside it. In a ventilated Class 2 Safe Room, air is drawn from outside but at a flow rate too small to create a measurable differential pressure.

○ **Class 3**. This class has no air-filtering capability and is unventilated. It is a basic safe room that derives protection only by retained clean air within its tight enclosure. Use of the Class 3 Safe Room is commonly referred to as sheltering-in-place.

Table 3-1: Comparison of the Three General Classes of Toxic-agent Safe Rooms

Class	Protection	Cost	Advantages and Limitations
1. Ventilated and pressurized with filtered air	high	high	Protection has no time limits, but it provides no protection against some toxic chemicals of high vapor pressure.
2. Filtration with little or no pressurization	medium	medium	Unventilated Class 2 is protective against all gases, but protection diminishes with duration of exposure (and against non-filterable gases).
3. Unventilated, no filtration	low	low	Protective against all agents, but protection diminishes with time of exposure. Carbon dioxide buildup may limit time in the shelter.

The Class 1 Safe Room provides the highest level of protection for most chemicals, but the lowest level of protection for those chemicals that are not filterable. It is also the most expensive option. Its disadvantage is that it does not protect against a limited number of toxic gases that cannot be filtered by conventional gas filters/adsorbers.

Although the Class 2 Safe Room employs air filters, it does not prevent the infiltration of outdoor air driven by natural forces of wind and buoyancy. It therefore provides a lower level of protection than a Class 1. If exposed to an unfilterable gas, the unventilated Class 2 Safe Room retains a level of protection provided by the sealed enclosure. The unventilated Class 2 Safe Room would thus not have a complete loss of protection as could occur with the gas penetrating the filter of a Class 1 or Class 2 ventilated Safe Room.

The Class 3 Safe Room, with no air filtration, is the simplest and lowest in cost. It can be prepared with permanent sealing measures or with the quick application of expedient sealing techniques such as applying duct tape over the gap at the bottom of the door or over the bathroom exhaust fan grille. The disadvantage is that there is no intentional ventilation; therefore, this class of safe room cannot conform to ventilation requirements of other types of emergency shelters.

Most safe rooms are designed as standby systems; that is, certain actions must be taken to make them protective when a hazardous condition occurs or is expected. They do not provide protection on a continuous basis. Merely tightening a room or weatherizing a building does not increase the protection to the occupants. Making the safe room protective requires turning off fans, air conditioners, and combustion heaters as well as closing doors and windows. It may also involve closing off supply, return, or exhaust ducts or temporarily sealing them with duct tape. In a residence, taking these actions is relatively simple and can be done quickly. In an office building, doing so usually requires more time and planning, as there may be several switches for air-handling units and exhaust fans, which may be at diverse locations around the building.

Unventilated safe rooms have been widely used in sheltering-in-place to protect against accidental releases of industrial chemicals. Local authorities make the decision on whether to shelter-in-place or evacuate based on conditions, the likely duration of the hazard, and the time needed to evacuate. Although sheltering-in-place (i.e., use of an unventilated safe room) is applicable for relatively short durations, experience shows that it may be necessary for people to occupy safe rooms for longer periods as a precautionary measure.

The potential for safe room stays of longer duration make it important to consider human factors in designing and planning safe rooms. Human factors considerations include ventilation, environmental control, drinking water, toilets, lighting, and communications.

3.2 HOW AIR FILTRATION AFFECTS PROTECTION

The addition of air filtering improves the protection a safe room provides, although there are limitations as to what gases can be filtered. Ventilation with filtered air also removes the time constraints associated with unventilated shelters.

To protect against the many gases, vapors, and aerosols that could be released in an accident or terrorist act requires three different filtering processes. Mechanical filtration is most commonly used for aerosols; physical adsorption, for chemical agents of low vapor pressure; and chemisorption, for chemical agents of high vapor pressure. These three processes can be provided by a combination of two types of filters: the HEPA filter to remove aerosols and a high-efficiency gas adsorber with impregnated carbon to remove vapors and gases. A filter system for a safe room must contain at least one HEPA and one gas adsorber in series, with the HEPA normally placed first in the flow stream.

HEPA adequately removes all toxic aerosols, including sub-micron size biological agents. A gas adsorber works for most, but not all gases/vapors. Several of the common industrial gases, such as ammonia, are not removed by the best broad-spectrum impregnated carbon available.

To protect against highly toxic chemicals, a Class 1 system requires ultra high-efficiency filtration, at least 99.999 percent removal in a single pass. HEPA filters, which are defined as having at least 99.97 percent efficiency against the most penetrating particle size (about 0.3 micron), have efficiencies greater than 99.999 percent against aerosols of 1- to 10-micron size, the most likely size range for biological-agent aerosols.

With a filtration system drawing outside air, the level of protection the safe room provides is a function of the filter efficiency. With an unventilated Class 2 system, the level of protection is not affected as greatly by changes in filter efficiency. For example, increasing filter efficiency from 99 percent to 99.999 percent in an unventilated Class 2 system improves the protection factor by about 1 percent. The same change in a Class 1 system yields a protection factor 1,000 times higher.

All filters have limited service life. In operation, a gas adsorber loads as molecules fill the micropores of the carbon, and a HEPA filter loads with dust and other particles to increase the resistance to flow. The adsorber loses capacity for gases over time when exposed to the atmosphere, even if air is not flowing through it. The shelf life of an adsorber ranges from 5 to 10 years when the filters are hermetically sealed in a container. The service life of the adsorber varies with the operating environment and is generally less than 5 years. For this reason, filters intended for use in safe rooms in the home or office are typically designed to remain sealed in a metal canister until they are needed in an emergency. This hermetic sealing can ensure the filters retain full filtering capacity for 10 years or more, although most manufacturers do not warranty them for more than 10 years.

Commercial filter units that are designed for indoor air quality can be used in an unventilated Class 2 Safe Room. There are many different models available from several manufacturers; however, the filtering performance varies over a wide range. These filter units can be ceiling-mounted, duct-mounted, or free-standing floor or table units having both HEPA filters and adsorbers (usually an activated carbon and zeolite mix). The HEPA filter element provides protection against a biological agent and other solid aerosols such as tear gas, while the adsorber protects against gases and vapors.

3.3 SAFE ROOM CRITERIA

This section presents criteria for selecting or designing a safe room for protection against airborne toxic materials. Although the protective envelope can be defined as the whole building, a room within the building (i.e., a safe room) can provide a higher level of protection if it is tighter than the building as a whole and/ or the location of the room is less subject to wind or buoyancy forces that induce infiltration.

Any type of room can be used as a safe room if it meets the criteria listed below. In office buildings, safe rooms have been established in conference rooms, offices, stairwells, and other large common areas. In dwellings, safe rooms have been established in bedrooms, basements, and bathrooms. The criteria are as follows:

○ **Accessibility**. The safe room must be rapidly accessible to all people who are to be sheltered. It should be located so that it can be reached in minimum time with minimum outdoor travel. There are no specific requirements for the time to reach a safe room; however, moving to the safe room from the most distant point in the building should take less than 2 minutes. For maximum accessibility, the ideal safe room is one in which one spends a substantial portion of time during a normal day. The safe room should be accessible to persons with mobility, cognitive, or other disabilities. Appropriate use of stirs or ramps when shelters are located above or below grade must account for such occupants.

○ **Size**. The size criterion for the toxic-agent safe room is the same as tornado shelters. Per FEMA 361, the room should provide 5 square feet per standing adult, 6 square feet per seated adult, and 10 square feet per wheelchair user for occupancy of up to 2 hours.

○ Tightness. There is no specific criterion for air tightness. With doors closed, the safe room must have a low rate of air exchange between it and the outdoors or the adjacent indoor spaces. Rooms with few or no windows are preferable if the windows are of a type and condition that do not seal tightly (e.g., older sliders). The room must not have lay-in ceilings (suspended tile ceilings) unless there is a hard ceiling above. The room should have a minimum number of doors, and the doors should not have louvers unless they can be sealed quickly. The door undercut must be small enough to allow sealing with a door-sweep weather strip or expediently with duct tape.

○ **HVAC system**. The safe room must be isolated or capable of being isolated quickly from the HVAC system of the building. If the selected room is served by supply and return ducts, modifications or preparations must include a means of temporarily closing the ducts to the safe room. In the simplest form, this involves placing duct tape or contact paper over the supply, return, and exhaust grilles and turning off fans and air-handling units. If there is a window-type or through-the-wall air conditioner in the selected room, plastic sheeting and tape must be available to place over the inside of the window and/or air conditioner, which must be turned off when sheltering in the safe room.

○ **Ventilation**. For Class 1 Safe Rooms, 15 cfm per person is the desired ventilation rate; however, the minimum ventilation rate is 5 cfm per person if that rate is adequate for pressurization. Class 3 and unventilated Class 2 Safe Rooms are suitable only for short-duration use, not only because the low ventilation rate when occupied can cause carbon dioxide levels to rise, but also because protection diminishes as the time of exposure to the hazard increases.

○ **Location**. For unventilated shelters (Class 3 and some Class 2), there are three considerations for location within a building. First, relative to the prevailing wind, the safe room should be on the leeward side of a building. Second, if there is a toxic-materials storage or processing plant in the community, the safe room should be on the side opposite the plant. Third, an interior room is preferable to a room with exterior walls, if it meets criteria for size, tightness, and accessibility. For a low-rise building, there is no substantial advantage in a room on the higher floors, and a location should not be selected based on height above ground level if it increases the time required to reach the shelter in an emergency.

○ **Water and toilets**. Drinking water and a toilet(s) should be available to occupants of a safe room. This may involve the use of canned/bottled water and portable toilets. Toilet fixture allowance is presented in FEMA 361.

○ **Communications**. For sheltering situations initiated by local authorities, the safe room must contain a radio with which to receive emergency instructions for the termination of sheltering. A telephone or cell phone can be used to receive emergency instructions and to communicate with emergency management agencies. Electrical power and lighting are also required.

3.4 DESIGN AND INSTALLATION OF A TOXIC-AGENT SAFE ROOM

After the room or location for the safe room has been decided based on the criteria listed above, the first design decision is to determine the class of safe room. Design details for the three classes of safe rooms are presented below. A Class 3 Safe Room is the simplest in that it requires only a tight enclosure. It is presented first because the requirements of the tight enclosure are common to all three classes. The unventilated Class 2 Safe Room, which involves the simplest application of a filter unit, is presented next, and the Class 1 Safe Room, which involves a more complex application of a filter unit, is presented last.

3.4.1   Class 3 Safe Room

Features of the Class 3 Safe Room can be either permanent or expedient. Guidance for preparing the safe room is presented in four parts:

○ How to tighten the room before an emergency – to permanently seal unintentional openings

○ How to prepare for sealing the room in an emergency – to temporarily close intentional openings such as ducts, doors, and windows in an emergency

○ How to prepare for rapid deactivation of fans

○ How to accommodate the safe use of air conditioning or heating in protective mode

3.4.1.1 Tightening the Room

○ **Ceiling-to-wall juncture.** Typically, most leakage occurs through the wall-to-ceiling and wall-to-floor junctures, particularly if suspended lay-in ceilings are used without a hard ceiling or a well-sealed roof-wall juncture above the lay-in ceiling. If the selected room has only a lay-in ceiling between the living space and attic space, the ceiling should be replaced with one of gypsum wallboard or other monolithic ceiling configuration.

○ **Floor-to-wall juncture.** A baseboard often obscures leakage paths at the floor-to-wall juncture, and to seal these leakage paths may require sealing behind the baseboards. One approach is to temporarily remove the baseboards and apply foam sealant in the gap at the floor-to-wall juncture. The alternate approach is to use clear or paintable caulk to seal the top and bottom of baseboards and quarter rounds. If there are electric baseboard heaters, the heaters should be temporarily removed to seal the wiring penetrations and the gap at the floor-to-wall juncture.

○ **Penetrations.** Measures for reducing air leakage through penetrations are as follows:

○ Seal penetrations for pipes, conduits, ducts, and cables using caulk, foam sealants, or duct seal.

○ Place weather-stripping (including a door sweep) on the door(s) of the safe room. If the selected room is a bathroom, and there is a supply duct but no exhaust duct, the door sweep may be omitted because it would reduce the supply flow rate in normal use. In this case, duct tape can be used to seal the gap beneath the door temporarily in an emergency. If there is carpeting in the safe room, a door sweep may be more effective than tape. There may be louvers in the door for return airflow, but they should not be modified to ensure proper ventilation can be maintained in normal conditions. Door louvers should be expediently sealed as described in Section 3.4.1.2.

○ Windows that are old and/or in poor condition can allow substantial leakage; however, newer, non-operable windows are not likely to require any sealing. Window leakage can be measured using a blower door to determine whether window replacement or sealing measures are necessary. In some cases, the leakage of windows, such as poorly maintained sliders, can be reduced only by replacing them or by using expedient sealing measures such as taping plastic sheeting over them.

○ Expanding foam can be used to seal electrical outlets and switches. Also, ready-made outlet sealers can be used to seal gaps behind switches and outlets.

3.4.1.2 Preparing for Rapidly Sealing the Room. The selected safe room may have one or all of the following intentional openings, which are necessary for normal operation. The openings must be

---

sealed so that the safe room can be used in a toxic-materials emergency unless the HVAC system for the safe room is designed to safely operate in the protective mode (as described below).

○ Supply and return ducts

○ Exhaust fan

○ Door louvers

○ Window-type air conditioner or unit ventilator

○ Door undercut

It is neither practical or advisable to seal these openings beforehand if the room is one that has normal day-to-day use, in which case plans and preparations should be made for sealing them temporarily during rapid transition to the protective mode. The sealing capability can be either permanent or expedient.

○ **Permanent capabilities for rapid sealing.** There are two general approaches to closing the intentional openings in transition to the protective mode. The first is to use hinged covers mounted within the safe room. The second is to use automatic dampers, particularly in ducts for supply, return, and exhaust. Hinged covers can be custom-made of sheet metal or wood, as shown in Figure 3-1, to be attached above or beside the opening for all applications except the door periphery. A hinged cover provides the capability to seal vents rapidly. In a safe room having several openings to be sealed, use of hinged covers allows the sealing to be completed more quickly than use of tape and adhesive backed plastic material.

Figure 3-1
Hinged covers facilitate the rapid sealing of supply, return, or exhaust ducts in a safe room.
SOURCE: BATTELLE

○ **Temporary measures for rapid sealing.** For expedient sealing, a small kit of materials should be provided in the safe room, along with a written checklist of the sealing measures required specifically for the safe room. The following is an example of a checklist that applies to a bathroom. The sealing supplies are contact paper, precut to size, and 2-inch wide painter's tape. Alternately, duct tape and plastic sheeting can be used.

   ○ Cover the door louvers with adhesive backed film (contact paper) 18 inches by 12 inches in size.

   ○ Cover the gap beneath the door with a strip of 2-inch-wide tape.

   ○ Cover the exhaust fan grille with adhesive-backed film (18 inches by 18 inches).

   ○ Cover the supply grille with adhesive backed film (12 inches by 12 inches).

   ○ Pour water into the floor drain and drains of the shower and sink to ensure the traps are filled.

In a bathroom, drain traps for the sink, tub, shower, or floor are usually filled with water, but should be checked to ensure water is present to prevent air leakage through the drain pipes.

A window-type or through-the-wall air conditioner can be sealed by turning it off, and placing either contact paper or plastic sheeting with duct tape over the air conditioner.

3.4.1.3 Preparing for Deactivation of Fans. Some safe room systems have been designed with the capability to automatically deactivate all fans in the building with a single switch. This single-switch control can also be designed to close dampers in outside-air ducts serving the safe room. The low-cost alternative to automatic fan shutoff is to record on a checklist the location of switches for all fans in the building, not just those that serve the safe room. This includes air-handling units, exhaust fans, supply fans, window air conditioners, and combustion heaters.

This checklist must also include the procedures for the purging step of sheltering-in-place (e.g., opening windows and doors, and turning on fans and air handlers that were turned off to shelter-in-place after the hazardous condition has passed).

3.4.1.4 Accommodating Air Conditioning and Heating. Conventional air conditioning and heating systems must not be operated in the protective mode because the fans directly or indirectly introduce outside air. This includes the air-handling units and fans serving spaces outside the safe room. An exception is combustion heaters of hydronic systems that are located in separately ventilated mechanical rooms.

In extreme weather conditions, however, confining people in a sealed room without air conditioning or heating can result in intolerable conditions, causing people to exit the safe room before it is safe to do so.

The mechanical ventilation system often has a higher potential for indoor-outdoor air exchange than the leakage paths of the enclosure subjected to wind and buoyancy pressures. Window-type air conditioners and unit ventilators cannot be used in the protective mode, because they introduce outdoor air, even when set to the recirculating mode. The dampers for outside air in such units seal poorly even when well maintained.

The following are options for air conditioning and heating systems that can be safely operated in the protective mode:

○ **Ductless mini split-type air-conditioner.** This type of room air conditioner, an alternative to the standard unitary window air conditioner, circulates air across the indoor coils without ducts and does not introduce outdoor air in either the normal or protective mode. The only required penetration through the safe room boundary wall is for a conduit for refrigerant tubing, suction tubing, condensate drain, and power cable.

○ **Electric heater or steam radiator.** Similar to the ductless split-type air conditioner, the electric or steam heater does not introduce outdoor air and does not require ducts.

○ **Fully enclosed air-handling unit.** An air-handling unit can be operated in a safe room in the protective mode only if the unit and its ducts are fully within the safe room (i.e., the unit is in an interior mechanical closet and the return ducts are not above the ceiling, beneath the floor, or outside the walls). If the air-handling unit draws outdoor air through a duct, it must also have a damper system for reliably cutting off outside air in the protective mode. This may require a set of three dampers: two dampers in the outside air duct with a relief damper between them that opens (to protected space) when the other two close. The air-handling unit must serve the safe room exclusively.

○ **Makeup air unit.** This is a once-through type unit for introducing fresh air; it is not applicable to an unventilated safe room. The makeup-air unit does not recirculate air through ducts; it supplies filtered air through duct coils for cooling and heating.

3.4.1.5 Safety Equipment. Unventilated safe rooms, whether Class 2 or Class 3, must have a carbon dioxide detector or monitor in the safe room.

3.4.2   Class 2 Safe Room

The design details of the enclosure presented above apply also to the Class 2 Safe Room, ventilated and unventilated. The ventilated Class 2 Safe Room is one that supplies filtered air from outside the safe room, but has inadequate air flow to pressurize the room. For the unventilated Class 2 Safe Room, the improvement in protection over the Class 3 Safe Room is determined by the flow rate and the efficiency of the particulate filter for aerosols and the efficiency of the adsorber for gases and vapors. These filter units, commonly referred to as indoor air purifiers, indoor air cleaners, or indoor air quality units, recirculate air within the safe room. There are four configurations:

○ Free-standing table top unit

○ Free-standing floor unit

○ Ceiling-mounted unit

○ Duct-mounted unit (with ducts completely inside the safe room)

3.4.2.1 Filter Unit Requirements for the Unventilated Class 2 Safe Room. The protection provided by an unventilated Class 2 Safe Room is determined by the clean-air delivery rate of the filter unit and the tightness of the enclosure. The clean-air delivery rate is a product of the filter removal efficiency (expressed as a decimal fraction) and the actual flow rate of the filter unit. If a high-efficiency filter unit is used, the clean-air delivery rate approaches the actual flow rate of the unit. If the filter has a single-pass efficiency of 50 percent, for example, the clean-air delivery rate is half the actual flow rate. For a given unit, the clean-air delivery rate is likely to be higher for aerosols than for gases and vapors because efficiencies of adsorbers are typically lower than the efficiencies of particulate filters in these units.

Many models of these indoor air purifiers are available commercially, but not all of them have performance suitable for use in protecting against toxic aerosols, gases, and vapors. The following are criteria for selection of recirculating filter units for use in safe rooms:

○ The filter unit must have both an adsorber containing activated carbon and a particulate filter.

○ The adsorber must have at least 1 pound of activated carbon for each 20 cfm of flow rate. For example, a 200-cfm unit requires at least 10 pounds of carbon adsorbent.

○ The particulate filter must have an efficiency of at least 99 percent against 1-micron particles.

○ The unit(s) must provide a total clean-air delivery rate of at least 1 cfm per square foot of floor area.

○ The adsorber must have the capability for chemisorption (i.e., for removal of gases that are not removed by physical adsorption).

There are also ventilated Class 2 Safe Rooms and essentially these are ones for which the filter unit has inadequate capacity to produce a measurable overpressure with the size of the selected safe room. In essence, the filter units are over-rated by the filter unit manufacturer. Generally, if a filter unit capacity in cfm is less than one-fourth the area (in square feet) of the selected safe room, depending on the type of construction, it will not produce a measurable overpressure. Matching the filter unit capacity to safe room size for Class 1 (pressurized) Safe Rooms is addressed in Section 3.4.3.2.

3.4.2.2 Installation and Operation. For the unventilated safe room, floor/table model filter units and ceiling-mounted models should be placed in the center of the room to maximize air mixing. There should be no obstruction to the airflow into and out of the filter units.

Duct-mounted models must conform to the requirements stated above for air-handling units. Ducts cannot be outside the envelope formed by the walls, ceiling, and floor.

The adsorbers of these commercial units are generally lacking in capability for filtering a broad range of high-vapor-pressure agents. Several of the common industrial chemicals (e.g., ammonia) are not removed.

The filter unit can be used routinely for indoor air quality; a spare set of filters should be kept on hand for use in a toxic-materials emergency, along with instructions for changing the filters so that the change can be made rapidly.

3.4.3   Class 1 Safe Room

Designing and installing a ventilated safe room is much more complex than an unventilated safe room, particularly with regard to the filter unit. Pressurization requires introducing air from outside the protective enclosure; therefore, the removal efficiency of the filters is more critical in determining the protection provided. The system must employ ultra-high efficiency filters, and it must allow no air to bypass the filter as it is forced into the safe room.

Except for military standards, there are no performance standards specifically for ultra-high efficiency adsorbers intended for protection of people from highly toxic chemicals. Performance of HEPA filters for aerosols is defined by ASME AG-1, *Code on Nuclear Air and Gas Treatment*, and N509, *Nuclear Power Plant Air-Cleaning Units and Components*. The specifications for filter units available commercially may present information that only partially defines the performance of an adsorber.

3.4.3.1 Selecting a Filter Unit for a Class 1 Safe Room. Generally, filter units available commercially are not designed to standards that ensure protection against highly toxic chemical, biological, and radiological materials. Some may provide very little protection, particularly if the manufacturer is not experienced in designing and building ultra-high efficiency filter units. Minimum requirements for the Class 1 applications are listed below. In purchasing a filter unit, certifications relative to the following requirements should be provided by the vendor:

○ The filter unit must have both a HEPA filter and an ultra-high-efficiency gas adsorber in series.

○ The adsorber must contain carbon impregnated ASZM-TEDA or the equivalent. Carbon mesh size should be 12x30 or 8x16.

○ The adsorber must have efficiency of at least 99.999 percent for physically adsorbed chemical agents and 99.9 percent for chemisorbed agents.

○ The adsorber must have a total capacity of 300,000 milligram (mg)-minutes per cubic meter for physically adsorbed chemical agents.

○ Bypass at the seals between the adsorber and its housing must not exceed 0.1 percent.

○ For installation of the filter unit outside the safe room, the fan must be upstream of the filters (blow-through configuration). For installation inside with a duct from the wall to the filter unit, the fan must be downstream of the filters (draw-through configuration).

○ If a flexible duct is used outside the shelter to convey air from the filter unit to the safe room, it must be made of a material resistant to the penetration of toxic chemicals.

○ If chemical manufacturing and storage facilities in the community present a special risk for release of toxic materials, special sorbents or sorbent layers may be required. In some cases, the chemicals produced/stored may not be filterable with a broad-spectrum impregnated carbon. For example, a nearby ammonia plant requires a special adsorber for protection against ammonia.

3.4.3.2 Sizing the Filter Unit for Pressurization. If a filter unit is undersized (i.e., it provides inadequate flow for pressurization), the result is substantially lower protection factors and the system becomes a ventilated Class 2 Safe Room. Filter unit(s) must be sized to provide makeup air at a flow rate sufficient to produce a pressure of at least 0.1 inch water gauge (iwg) in the shelter for protected zones of one or two stories. Taller buildings require an internal pressure higher than 0.1 iwg to overcome the buoyancy pressures that result in extreme weather conditions (i.e., large temperature differences between the inside and outside of the safe room).

The airflow rate needed to achieve this pressure in a safe room varies with the size and construction of the safe room. Generally, commercial filter units designed for home or office safe rooms are under-rated with regard to the quantity of air needed for pressurization. For safe rooms of frame construction and standard ceiling height, most can be pressurized to 0.1 iwg with airflow in the range of 0.5 to 1 cfm per square foot. Table 3-2 provides additional guidance in estimating the size of a filter unit for a safe room based on square footage.

The recommended procedure for ensuring that pressurization can be achieved is to perform a blower door test after all permanent sealing measures have been completed. The test should be conducted per ASTM E779-03, *Standard Test Method for Determining Air Leakage by Fan Pressurization* with temporary sealing measures in place.

Table 3-2: Leakage per Square Foot for 0.1 lwg (estimated makeup airflow rate per square foot (floor area) to achieve an overpressure of 0.1 inch water gauge)

Construction Type	cfm per square foot of floor area
Very tight: 26-inch thick concrete walls and roof with no windows	0.04
Tight: 12-inch thick concrete or block walls and roof with tight windows and multiple, sealed penetrations	0.20
Typical: 12-inch thick concrete or block walls with gypsum wall board ceilings or composition roof and multiple, sealed penetrations	0.50
Loose: Wood-frame construction without special sealing measures	1.00

3.4.3.3 Other Considerations for Design of a Class 1 Safe Room

○ **Heating and cooling the safe room**. A safe room does not require heating and cooling; however, in extreme weather, the conditions in the safe room may become uncomfortable due to the lack of ventilation or the introduction of outdoor air that is not tempered. In hot weather, this can be worsened by the temperature rise that occurs as air passes through the filter unit. Because of the relatively high pressure drop across the high efficiency filters, the temperature of the air typically increases by 5 to 10 degrees Fahrenheit as it passes through the filter unit. The use of inefficient fans, such as brush-type high-speed fans, should be avoided for this reason, because a temperature rise of 15 degrees can result.

○ **Control system**. An interlocking system should be considered for closing automatic dampers (as shown in Figure 3-2), turning off air-handling units, exhaust fans, and ventilation fans serving the building's unprotected spaces while the safe room is in the protective mode. This increases the level of protection the safe room provides against an outdoor release of agent.

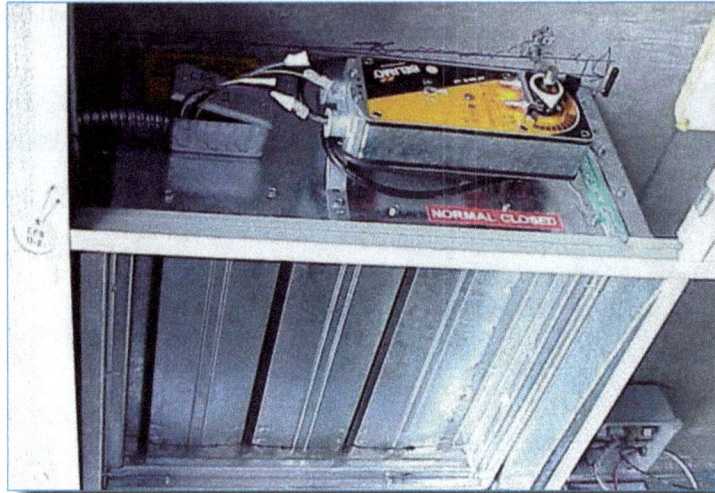

Figure 3-2

Automatic dampers are used to isolate the safe room from the ducts or vents used in normal HVAC system operation.

SOURCE: CHEMICAL STOCKPILE EMERGENCY PREPAREDNESS PROGRAM

○ **Heating system safety**. If a fuel-fired indirect heater (i.e., heat exchanger) is used to heat the safe room, a carbon monoxide detector with a visual display and an audible alarm should be installed in the safe room. Electric coil and hot-water coil systems do not require a carbon monoxide detector.

○ **Pressure gauge**. For Class 1 Safe Rooms, the pressure gauge is the indicator that the system is operating properly. This gauge displays the pressure in the safe room relative to outdoors or outside the safe room indoors. If the reference pressure is measured indoors, the readings can be subject to variations caused by fan pressures unless other building heating, ventilation, and air conditioning (HVAC) fans are turned off when the safe room is in use. Reading the reference pressure outdoors can be subject to positive and negative variations caused by air flows over and around the building. If the pressure sensor is outdoors, it should be shielded from the wind. Indoors is the best location if the building HVAC fans are turned off when the safe room is in use.

3.5 OPERATIONS AND MAINTENANCE

For a shelter to be successful, it is critical to have an understanding and dedication to operations and maintenance. Depending on the shelter type, specific operations instructions and maintenance are needed.

○ **Instructions and checklists**. As a minimum for operating procedures, the condensed operating and maintenance instructions should be posted in each safe room. The operating instructions should explain the steps of placing the safe room into operation and may be as simple as a one-page typed checklist; instructions that should be included are safe room operating procedures, a list of doors to be secured, a list of switches for fans to be turned off, stations/channels for emergency instructions, emergency phone numbers, and dates by which filters should be changed, if applicable.

○ **Status indicators**. For safe rooms that require multiple automatic dampers to isolate the safe room from the HVAC ducts in the protective mode, status lights and/or visual indicators should be used to show the position of each damper. Indicators can also be used to show door position, if there are multiple boundary doors in the safe room. Each status light should be marked with a reference number corresponding to a diagram so an operator can easily determine the location of any damper/door and conduct troubleshooting if problems occur. The indicator lights should have push-to-test capability for the light bulbs of the status lights.

○ **Public-address system**. For safe rooms in large buildings, a public address system is the most efficient means of instructing building occupants to proceed to a safe room in an emergency. Telephone or audible alarm systems can also be used, but they are less efficient than a broadcast voice system. Communications systems (telephone, alarm, and mass notification systems) should be tied to emergency phone systems. Non-verbal warning systems are generally less effective because they require training on the meaning of different types of alarm sounds.

○ **Auxiliary or Battery Power**. Class 3 Safe Rooms do not require electrical power to protect their occupants. Class 1 and Class 2 Safe Rooms require power for the air-filtration units to protect at a higher level than Class 3. If power is lost in a Class 1 or Class 2 Safe Room, it will continue to protect at the level of a Class 3 Safe Room as long as the room remains sealed. Power failure, therefore, does not lead to protective failure, but rather a reduced level of protection and reduced level of comfort in some conditions. For this reason, auxiliary power is not essential for a CBR safe room. Auxiliary power is provided on some CBR safe rooms so that the highest level of protection and comfortable conditions can be maintained if a power loss is caused by or coincides with the event causing the release of toxic agent.

3.5.1   Operating a Safe Room in a Home

The essence of operating a Class 3 Safe Room is to close the safe room and ensure that building fans, combustion heaters, and air conditioners are turned off so that they do not cause an exchange of air between the safe room and its surroundings. General procedures for the home safe room are as follows:

○ Close all windows and doors (both interior doors and exterior doors of the home).

○ Turn off the central fan, exhaust fans, window air conditioners, or combustion heaters in the home.

○ Enter the safe room. If there is no telephone in the safe room, take a cell phone or portable phone into the safe room for emergency communications.

○ Close the safe room door and apply tape to the periphery of the door, unless there are weather seals on the door.

○ Turn on a radio or TV in the safe room and listen for emergency information.

○ If the safe room has a carbon dioxide detector, monitor it, particularly if the time in the sealed safe room exceeds 1 hour.

○ When the "all clear" determination is made, open the windows and doors, turn on ventilation systems, and go outside until the house has been fully aerated.

The following is a list of supplies for the safe room:

○ Rolls of duct tape for sealing doors and securing plastic over vents and windows

○ Pre-cut plastic sheeting to fit over supply and return vents (also for windows if they are judged to be less than airtight)

○ Battery operated radio with spare batteries

○ Flashlight with spare batteries

○ Drinking water

○ First aid kit

○ Telephone (cell phone) for emergency instructions

3.5.2 Operating a Safe Room in an Office Building

For the office-building safe room, the supplies are generally the same as the home safe room listed above.

Procedures differ in that there are likely to be more exterior doors to be closed and multiple locations for the switches that control building fans. To ensure that all doors are closed and fans are turned off, an emergency plan and checklist should be developed, assigning employees to these tasks at various locations in the building. The general procedures for the office building safe room are as follows:

○ A building-wide announcement is made for all building occupants to proceed to the designated safe room(s).

○ Assigned monitors secure all exterior doors and windows.

○ Assigned building engineering staff turn off all air handling units, ventilation fans, and window air conditioners as applicable (or security turns them off if a single switch capability has been installed).

○ Safe room doors are secured as soon as possible once all who are assigned to the safe room have entered.

○ Employees and visitors are accounted for by use of a roster and visitor's sign-in sheets.

○ Emergency information is obtained by radio, TV, telephone, or cell phone in the safe room.

○ If the safe room has a carbon dioxide detector, monitor it if the time in the safe room exceeds 1 hour.

○ When the "all clear" determination is made, open the windows and doors, turn on ventilation systems, and go outside until the building has been fully aerated.

3.5.3   Operating Procedures for a Class 1 Safe Room

Operating procedures for Class 1 (pressurized) Safe Rooms are similar to those of Classes 2 and 3.

The system is turned on immediately upon receipt of a warning.

Control panels for Class 1 systems typically include pressure gauges and status lights for automatic dampers, which provide assurance that the system is operating properly and a means of troubleshooting if the system does not pressurize.

Tape, plastic, and carbon dioxide detectors are not necessary in the Class 1 Safe Room.

3.6 MAINTAINING THE CBR SHELTER

Depending on the shelter type, the shelter may require more maintenance. Shelters that have increasing levels of protection from filters will require more frequent checks and will require more funding to keep them operational.

3.6.1 Maintenance for a Class 3 Safe Room

The Class 3 Safe Room has no air filtration equipment and, therefore, requires little or no routine maintenance. It has no mechanical equipment unless there are dampers for isolating the air conditioner (configured for fail-safe operation). Maintenance requirements are limited to periodically checking supplies for deterioration or loss: duct tape, plastic sheeting, radio spare batteries, flashlight spare batteries, drinking water, and first aid kit.

3.6.2 Maintenance for a Class 2 Safe Room

The filter unit used in a Class 2 safe room is an indoor air quality filter unit (see Figure 3-3) and, as such, it can be used routinely to improve the air quality in the spaces in or around the designated safe room. If this is done, a spare filter set, both adsorber and HEPA filter, should be stored in a sealed bag in the safe room along with instructions and any tools needed for changing the filter quickly in an emergency. Other supplies to be checked on a regular basis are the same as listed for the Class 3 Safe Room above.

Figure 3-3
A tabletop recirculation filter unit with a substantial adsorber is a simple means of providing higher levels of CBR protection to unventilated safe rooms.

SOURCE: BATTELLE

3.6.3   Maintenance for a Class 1 Safe Room

Maintenance of the Class 1 Safe Room consists primarily of serviceability checks and replacing filters. Serviceability checks should be performed about every 2 months by turning the system on and checking for the following while it is operating:

○ **System pressure**. The system pressure is indicated by a gauge typically mounted on the control panel, with the correct operating range marked on the gauge. If the pressure is outside this range while the system operates, troubleshooting should be initiated.

○ **Isolation dampers**. Correct damper positioning is indicated by damper status lights on the control panel. Troubleshooting should be initiated if the status lights indicate a damper is not properly positioned.

○ **Relief damper**. If the system contains a pressure-relief damper, it should be visually inspected while the system is operating. A properly functioning relief damper should be open when the safe room is pressurized, and it should close immediately when a door is opened into the safe room, releasing pressure.

○ **HEPA filter resistance**. The differential pressure across the HEPA filter is measured by a gauge mounted on the filter unit with taps on either side of the HEPA filter. If the pressure across the filter is greater than specified (approximately 3 iwg or higher), it is an indication that the HEPA filter has become loaded with dust and its higher resistance is reducing the flow rate of the filter unit. If such is the case, the HEPA filter should be changed.

○ **Cooling system**. If the safe room supply air is cooled and heated, the temperature of the air flowing from the supply register should be checked with a thermometer during serviceability checks. In warm weather, this should be approximately 55 degrees if the cooling system is operating properly.

○ **Door latches**. All doors into the safe room should be adjusted to latch automatically with the force of the door closer. For safe rooms with multiple doors, leakage past unlatched doors can cause internal pressure to fall below the specified operating range.

○ **Weather stripping**. The weather stripping on each door on the boundary of the safe room should be visually inspected to ensure it has not been removed or damaged through wear and tear. For wipe seals at the bottom of the door, the alignment and height of the seal above the floor should be inspected and adjusted as necessary.

○ **Filters**. Routine maintenance includes replacing filters. If a canister-type filter is used, it is replaced as a unit at its expiration date. For other types of filter units, three types of filters are replaced: the pre-filter, HEPA filter, and carbon adsorber. Ideally, with only intermittent operation, all three types of filters should be replaced at the same time, every 3 to 4 years. This period is defined mainly by the service life of the adsorber.

Each time the CBR filters are replaced, in-place leakage testing should be performed, except in the case of canister filters (see Figure 3-4), to ensure the critical seals between the filters and/or between the mounting frame and the filters are established properly (i.e., there is no leakage past the filters' peripheral seals). To test the seals of the HEPA filter, the unit is challenged with an aerosol; poly-alpha olefin (PAO) is the industry standard. To test the seals of the adsorbers requires a chemical that is loosely adsorbed in the filter bed. Halide gases are typically used for this purpose. For the adsorber, the criterion is that the leak must be less than 0.1 percent of the upstream concentration. For the HEPA filter, the criterion is 0.03 percent. Procedures for both tests are described in American National Standards Institute/American Society of Mechanical Engineers (ANSI/ASME) N510, *Testing of Nuclear Air Treatment Systems.*

Figure 3-4
A canister-type filter unit is often used for Class 1 Safe Rooms to maximize storage life of the filters.

SOURCE: BATTELLE

3.7 COMMISSIONING A CLASS 1 CBR SAFE ROOM

Commissioning applies to the Class 1 Safe Room. It involves testing, checking the configuration, and performing functional checks to ensure the safe room has been installed properly, protects as intended, and can be operated and maintained by its owner. Commissioning addresses not only the safe room and its components, but also the operations and maintenance instructions.

For a Class 1 Safe Room, the principal performance indicator is the pressure developed in the safe room by the flow of filtered air. Commissioning requires measuring the internal pressure, the supply air flow rate, and leakage at the seals of the filters. If an air conditioning and/or heating system and dampers are part of the system, it also requires verifying their proper function. The following should be addressed in commissioning a Class 1 Safe Room:

3.7.1   Measurements

○ Measure the flow rate of filtered air and compare with design flow rate. Airflow rate measurements are usually made by certified test-and-balance contractors. A filter unit with integral motor-blower and fixed supply-duct length may not require airflow measurements after installation.

○ Conduct in-place leakage testing to determine if filters are sealed properly to their mounting frames to prevent air bypassing the filters. This is necessary if the filter unit has replaceable filters. If a canister type filter unit is employed, these critical seals are factory tested, and in-place testing for bypass is not necessary.

○ Measure the pressure in the safe room with a calibrated gauge independent of the installed pressure gauge.

○ Measure the temperature of the supply air and compare it with design values for both heating and cooling modes.

3.7.2   Configuration

○ Visually inspect the seals applied to wall penetrations (pipes, cables, conduit) and to doors (weather-stripping and wipe seals).

○ For filter units with replaceable filters, determine that the filter unit has been installed with adequate clearance for changing the filters.

○ Verify that the pressure-sensing tubes for the pressure gauge have been installed properly, reference pressure sensors have been appropriately placed to provide accurate ambient pressure readings without the effects of dynamic pressure, and that they are shielded to prevent blockage by moisture, insects, etc.

○ If there is air conditioning or heating, inspect to determine that outside air will not be drawn in through closed dampers or other leakage points.

○ Verify that gauge and status lights of the control panel gauges are marked with operating ranges.

○ Verify that markings, signs, and condensed operating instructions are adequate for the user to operate the system properly.

3.7.3   Functionality

○ Visually inspect all dampers to ensure they move freely and assume the correct position for both normal and protective modes.

○ For a mechanical pressure gauge, determine that the gauge has been zeroed properly.

○ If there is a low-pressure alarm or status indicator, verify it has been adjusted to the correct pressure threshold.

○ Verify that status lights accurately indicate position/operation of dampers and fans.

○ Verify the push-to-test capability for status lights.

○ Verify that instructions for operating and maintaining the system are available at the safe room and provide clear and accurate guidance for an untrained operator to activate the protective system.

○ Verify the operation of communications equipment for safe room occupants.

○ Verify the proper function of the change-HEPA gauge or indicator.

3.8 UPGRADING A CBR SAFE ROOM

The simplest and least costly upgrade in protective capability for a safe room is to upgrade from Class 3 to unventilated Class 2. This involves adding a recirculation filter unit, the simplest of which is a free-standing unit. This type of filter unit is available in many models, flow rates, filter types, and cost levels; however, not all have the same capability. Requirements for the grade of adsorbers and particulate filters, as well as flow rate per square foot of safe room, are listed in Section 3.4.2.1.

Upgrading from Class 2 or 3 to Class 1 involves the greatest expense. It can be as simple as purchasing a high-efficiency filter unit and installing it to supply filtered air through a special duct. Rules of thumb on flow rates required to pressurize the safe room are presented in Section 3.4.3.2. An interlocking system should be considered for turning off air-handling units, exhaust fans, and ventilation fans of the building's unprotected zones while the safe room is in the protective mode.

3.9 TRAINING ON THE USE OF A SAFE ROOM

As is the case with fire safety, all people who are to be protected in a safe room, whether in a home or commercial building, must be familiar with the procedures of using the safe room. Getting into it quickly and closing the safe room door(s) is important for all types of shelters. Whether sheltering from toxic agents, blast, or storms, the protection a safe room provides is compromised when it is not fully closed.

Training the people who work or reside in the building on safe room procedures has four main objectives:

○ To familiarize them with the locations of the safe rooms and the procedures for using them.

○ To inform them about who the emergency manager/coordinator of the building is, what his/her responsibilities are, and how he/she can be contacted.

○ To develop an understanding of the range of protective responses, including evacuation, and what to do for each of the possible protective actions.

○ To develop an employee awareness of the threats and hazards. Trained building occupants can serve to detect threats and reduce the time to respond by being aware of indications of suspicious activities, symptoms of toxic agent exposure, or odors from chemical releases.

Plans should be made to conduct a safe room drill, similar to a fire drill, semi-annually.

3.10 CASE STUDY: CLASS 1 SAFE ROOM

This case study describes a Class 1 Safe Room, one that is ventilated and pressurized with an ultra-high efficiency filter unit in a multi-story office building.

The concept of operation for this safe room is that, in response to a release of toxic chemicals outside the building, security personnel activate the safe room filtration unit and turn off all other HVAC fans in the building to place the building in the shelter-in-place mode. Employees are instructed via the public-address system to proceed immediately to the safe room. They remain in the safe room until building security officers, consulting with the local emergency management agency, determine that there is no longer a hazard.

A ventilated, pressurized safe room was selected so that a large number of people could be protected during a sheltering period of 1 to 2 hours without the potential for carbon dioxide buildup.

Task 1. Select the Safe Room Space

Applying the criteria of adequate space, accessibility, and capability to be rapidly secured, the fire-rated stairwell is the best choice in a multi-story office building for a safe room.

The safe room must accommodate all occupants and visitors in this portion of the building. The stairwell is 11 stories high (140 feet) with cross-section dimensions of 35 feet by 12 feet. At 5 square feet per person, it can accommodate 600 people.

The stairwell meets the criterion of accessibility because it spans all floors of the building and is accessible to all building occupants in an estimated 1 minute or less. It also provides access for people with impaired mobility.

As a fire-rated stairwell, it is unventilated. It can therefore be rapidly secured and isolated from the mechanical ventilation system and the other spaces of the building. With no mechanical ventilation, it can be secured by simply closing its doors, which are normally held closed by door closers for fire-safety purposes.

Task 2. Determine How Well the Selected Space Can Be Sealed

Airtightness is indicated by the construction of the stairwell, which is of cast-in-place concrete with few wall penetrations, none of which has apparent sealing requirements beyond caulk or foam sealant. Penetrations include sprinkler standpipes and conduit/ cables for wall-mounted lights at each landing, an emergency telephone, and intercom. The doors into the stairwell are fire-rated and have an undercut of ½ to 1 inch. Collectively, the undercut of the doors into the stairwell is the largest leakage path to be sealed.

To confirm the airtightness and estimate the airflow capacity of the filter unit, a blower-door test, as illustrated in Figure 3-5, was performed per ASTM E779-03. Use of the blower door in the depressurization mode can also facilitate finding leaks that may not be readily apparent.

Figure 3-5
A blower door test on the selected safe room aids in estimating the size of air-filtration unit required and in identifying air leakage paths.
SOURCE: BATTELLE

Task 3. Determine the Level of Safe Room Positive Pressure Required

For a safe room of this height, buoyancy pressure is significant, and both wind pressure and buoyancy pressure must be considered in determining the safe room operating pressure. As a corner stairwell, it has two exterior walls. The pressure requirement is defined as the velocity pressure of a 20-mph wind plus the buoyancy pressure that occurs against ground-level doors and other points of leakage at winter design conditions. With a height of 140 feet, the maximum buoyancy pressure at the lowest level of the stairwell is calculated at 0.11 iwg for a 60-degree Fahrenheit indoor-outdoor temperature differential. Adding the maximum wind pressure of 0.2 iwg for a 20-mph wind yields a design pressure of 0.3 iwg. At an internal pressure of 0.3 iwg, the force required to open the doors into the stairwell is slightly less than 30 pounds at the door handle for the (inward opening) doors, the maximum door opening force allowed by fire code.

Ideally, a blower door-test is performed after the permanent sealing measures, including door seals, have been added. In performing the blower door test before these permanent sealing measures are applied, the doors are taped temporarily at the periphery to simulate their being permanently sealed. Results of the blower door test, graphed in Figure 3-6, show that, with doors sealed with tape, the stairwell could be pressurized to 0.3 iwg with about 3,000 cubic feet per minute (cfm).

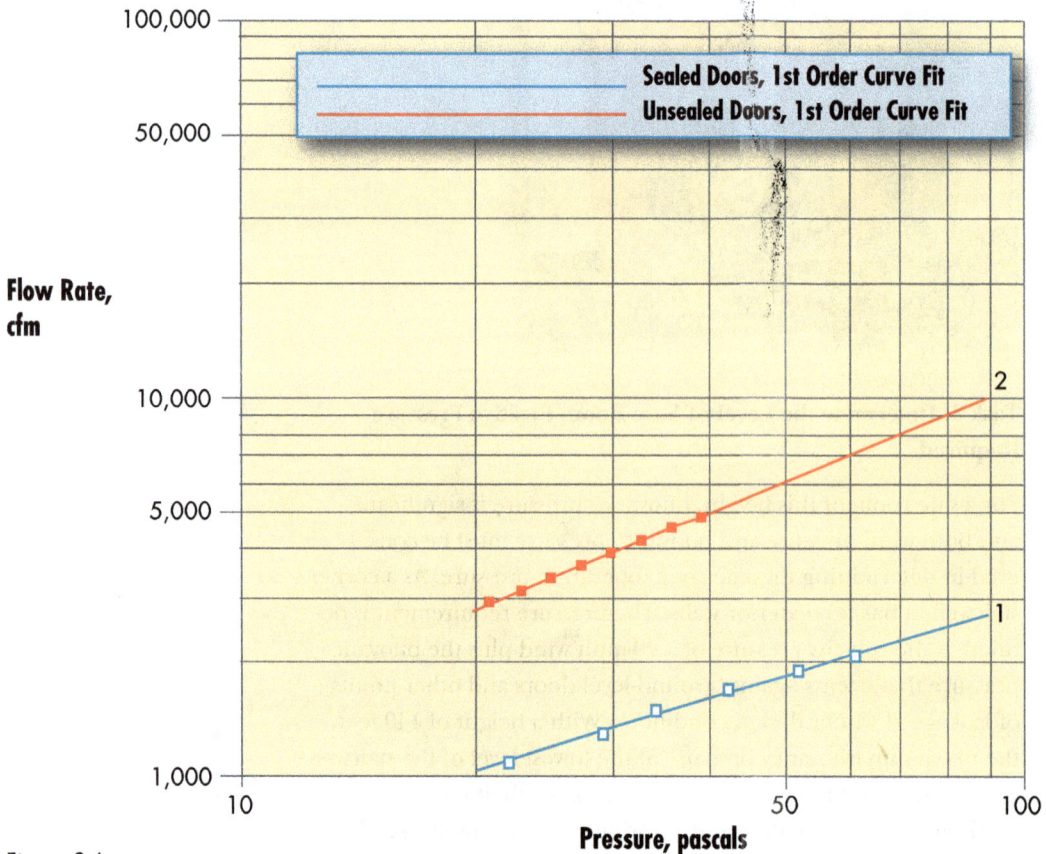

Figure 3-6
Blower-door test results on the stairwell selected for a safe room
SOURCE: BATTELLE

CBR THREAT PROTECTION

Task 4. Determine How Much Filtered Air Flow is Required

The total filtered, ventilation air flow required is the larger of: (1) the flow rate required for pressurization and (2) the flow rate required to supply greater than 5 cfm of outside air per person in the safe room. At a maximum safe room occupancy of 600 people, the 5 cfm/person ventilation requirement is met with a flow rate of 3,000 cfm. Using a filter unit of 4,000 cfm capacity yields a ventilation rate greater than 5 cfm per person. This ventilation flow rate is not adequate; however, to deliver the cooling required for the heat load of the fully occupied stairwell, two fan-coil units are added to the lower levels for additional cooling.

Task 5. Design and Install the System

Design and installation requires a licensed mechanical contractor with experience in the installation of high-efficiency filtration systems.

The safe room requires the following components installed in the stairwell and a penthouse mechanical room adjacent to the stairwell:

- Air-filtration unit and supply fan of 4,000-cfm capacity
- Pre-heat coil cabinet containing pre-filters, heating coils, and isolation damper
- Cooling-coil cabinet containing coils and an isolation damper
- Spiral ductwork
- Control panel with pressure gauge and system status lights
- Remote on/off switch connected via the building automation system
- Pressure relief damper at the lowest level of the stairwell
- Supply register at the top level of the stairwell
- Pressure sensor and thermostats installed in the stairwell
- Weather stripping and wipe seals on doors into the stairwell

Air-filtration Unit Type. A filter unit employing an adsorber containing ASZM-TEDA carbon of 12x30 mesh size was selected to provide ultra-high efficiency filtration of a broad spectrum of toxic chemicals. A military radial-flow filter set, carbon adsorber, and HEPA filter, shown in Figure 3-7, were selected. Manufactured to a government purchase description, the 4,000-cfm filter unit employs 20 replaceable sets of radial-flow filters. The filter sets were purchased from Hunter Manufacturing Company, Solon, OH, part number HF-200S.

Figure 3-7
A military radial-flow CBR filter set was selected for safe room filtration.

PHOTO COURTESY OF HUNTER MANUFACTURING COMPANY

Air-filtration Unit Location. A penthouse mechanical room adjacent to the stairwell is selected as the mounting location for the filter unit to provide an elevated and secure location for the equipment and its air intake. This is selected to achieve physical security of the intake and to make it most-distant from ground level releases. The filter unit, with its access panel removed, is shown in Figure 3-8.

Isolation Dampers. Two sets of automatic dampers are installed in the system, one upstream and one downstream of the filter unit to isolate the filters when not in use.

Figure 3-8
A 4,000-cfm filter unit using radial flow filters was selected for the stairwell safe room.

SOURCE: BATTELLE

Pre-heat Coil Module. This module contains pre-filters, pre-heat coil, and an outside air damper. Pre-filters selected for the system are ASHRAE 25-35 percent pleated. The building in which this system is installed has chilled water and hot water service from a central plant.

Cooling-coil Module. Cooling coils and an isolation damper are contained in a module mounted on the wall between the mechanical room and stairwell. Temperature control of the pressurization air is maintained with an electronic thermostat located in the stairwell.

Control Panel. The control panel as shown in Figure 3-9 has the following controls and indicators:

○ Start/stop switch

○ Status lights for the system, supply fan, and two isolation dampers

○ Pressure gauge indicating the pressure differential between the stairwell and the adjoining hallway

Figure 3-9
The Class 1 Safe Room control panel has a system start/stop switch, status indicators for dampers, and a pressure gauge.
SOURCE: BATTELLE

Relief Damper. In supplying filtered air from the top of the stairwell, the carbon dioxide concentration in the occupied stairwell increases with the vertical distance from the source of fresh air, because leakage paths are evenly distributed along the vertical axis. To ensure carbon dioxide levels remain within safe limits at the lowest levels of the stairwell and to facilitate removal of heat and humidity generated by the occupants, a relief damper was installed at the lowest level to maximize the flow-path length for clean air. The relief damper was adjusted to prevent the internal pressure from exceeding 0.30 iwg, a pressure above which the doors that open into the stairwell could require more than 30 pounds of force at the door handle to open (depending upon door closer force).

Door Seals. Weather-type seals were installed on doors into the stairwell to minimize air leakage around the closed doors. According to blower door test results, leakage around the stairwell doors before the addition of weather-stripping and wipe seals was substantial (approximately 3,000 cfm at 0.3 iwg).

Warning System. A public address system was installed in the building so that voice messages could be broadcast throughout the building to notify people at any location of an emergency.

Accessibility. The stairwell offers a point of entry accessible to wheelchairs, and each landing provides an area adequate for two wheelchairs without blocking access to the door.

Drinking Water. The stairwell has no accommodation for drinking water. There were plans to make water available by storing bottled water in a compartment on each landing of the stairwell.

Communications. Each level of the stairwell has an emergency telephone and an intercom to provide contact with the security operations center.

Cost. Total cost of the installed system was $190,000, or about $300 per person sheltered. This does not include the cost of the public address system or the single-switch fan shutdown for the building. Cost of the filter unit and initial set of filters was about one-fourth the total cost of the system. Costs included:

○ Purchase of filter unit and supply fan, $28,000

○ Purchase of CBR filters, 20 sets, $22,000

○ Detailed design and purchase and installation of other components, $140,000:

   ○ Chilled-water coils, hot water coils, piping, insulation, supports

   ○ Double-wall and single-wall spiral ductwork

   ○ Two isolation dampers and one relief damper

   ○ Control panel and remote activation capability

   ○ Concrete equipment pads

   ○ Door sweeps and jamb seals on stairwell doors

   ○ Sealing penetrations through stairwell walls

   ○ Motor control center and electrical service

   ○ Test, adjust, and balance

   ○ In-place leak testing of filter unit

   ○ Signage for condensed operating instructions

EMERGENCY MANAGEMENT CONSIDERATIONS   4

4.1 OVERVIEW

This chapter first outlines how DHS has planned for and responds to incidences of natural significance that would affect a shelter. These plans, policies, and procedures may be mirrored or modified by shelter owners and/or communities. Users of this document should check with local emergency management to determine their capabilities and plans for responding to an incident. The data presented for the Federal Government's approach to emergency management can be used to develop shelter operations plans and shelter maintenance plans.

4.2 NATIONAL EMERGENCY RESPONSE FRAMEWORK

On December 17, 2003, Homeland Security Presidential Directive (HSPD) 8: National Preparedness was issued. HSPD-8 defines preparedness as "*the existence of plans, procedures, policies, training, and equipment necessary at the Federal, State, and local level to maximize the ability to prevent, respond to, and recover from major events. The term 'readiness' is used interchangeably with preparedness.*" HSPD-8 refers to preparedness for major events as "*all-hazards preparedness.*" It defines major events as "*domestic terrorist attacks, major disasters, and other emergencies.*"

The Department of Homeland Security developed the National Response Plan (NRP) and the Catastrophic Supplement to the NRP and is now encouraging state and local government, private industry, and non-government organizations to achieve a multi-hazards capability as defined in the National Preparedness Goal.

The Homeland Security Digital Library (HSDL, https://www.hsdl.org) should be consulted for the following publications. The HSDL is the nation's premier collection of homeland security policy and strategy related documents.

The National Incident Management System

The National Incident Management System (NIMS) integrates existing best practices into a consistent, nationwide approach to domestic incident management that is applicable at all jurisdictional levels and across functional disciplines in an all-hazards context. https://www.hsdl.org/homesec/docs/dhs/nps14-030604-02.pdf

National Response Plan (Final) Base Plan and Appendices

The President directed the development of a new National Response Plan (NRP) to align Federal coordination structures, capabilities, and resources into a unified, all-discipline, and all-hazards approach to domestic incident management. https://www.hsdl.org/homesec/docs/dhs/nps08-010605-07.pdf

National Preparedness Goal [Final Draft]

The President directed the development of a National Preparedness Goal that reorients how the Federal government proposes to strengthen the preparedness of the United States to prevent, protect against, respond to, and recover from terrorist attacks, major disasters, and other emergencies. https://www.hsdl.org/homesec/docs/dhs/nps03-010306-02.pdf

National Response Plan

In Homeland Security Presidential Directive (HSPD)-5, the President directed the development of a new National Response Plan (NRP) to align Federal coordination of structures, capabilities, and resources into a unified, all discipline, and all-hazards approach to domestic incident management. This approach is unique and far-reaching in that it, for the first time, eliminates critical seams and ties together a complete spectrum of incident management activities to include the prevention of, preparedness for, response to, and recovery from terrorism, major natural disasters, and other major emergencies. The end result is vastly improved coordination among Federal, state, local, and tribal organizations to help save lives and protect communities by increasing the speed, effectiveness, and efficiency of incident management.

DHS has identified two emergency levels: routine and catastrophic, as shown in Figure 4-1. The types of emergencies
that occur on a daily basis, such as car accidents, road spills,
or house fires, are routine events. Catastrophic events, such as
tornadoes, terrorist attacks, or floods, tend to cover a larger
area, impact a greater number of citizens, cost more to recover
from, and occur less frequently. Emergencies are complicated
as the extent increases due to the additional layers of coordination and communication that need to occur as the event crosses
jurisdictional boundaries and overburdens the resources at the
origin of the event.

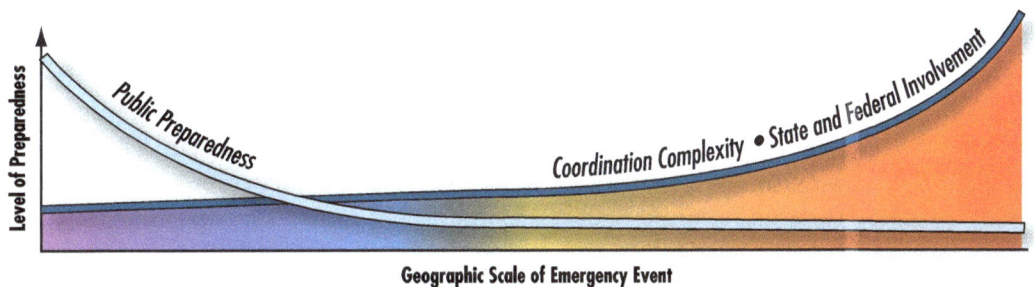

| Classification | Routine | | Catastrophic | |
	Local	Regional	State	National
Examples	• Minor Traffic Incidents • Minor Load Spills • Vehicle Fires • Minor Train/Bus Accidents • Accidents with Injuries but No Fatalities	• Train Derailment • Major Bus/Rail Transit Accidents • Major Truck Accidents • Multi-vehicle Crashes • HazMat Spills • Accidents with Injuries and Fatalities	• Train Crashes • Airplane Crashes • HazMat Incidents • Multi-vehicle Accidents • Tunnel Fires • Multiple Injuries and Fatalities • Port/Airport Incidents • Large Building Fire or Explosion • Industrial Incidents • Major Tunnel/Bridge Closure	• Terrorist Attack/WMD • Floods, Blizzards, Tornadoes • Transportation Infrastructure Collapse • Extended Power/Water Outages • Riots • Mass Casualties
Expected Event Duration	0 – 2 Hours	2 – 24 Hours	Days	Weeks

Figure 4-1   Preparedness versus scale of event

SOURCE: DHS NATIONAL GEOSPATIAL PREPAREDNESS NEEDS ASSESSMENT

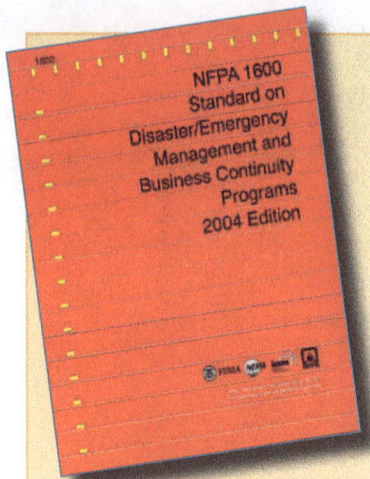

The National Fire Protection Association developed, in cooperation and coordination with representatives from FEMA, the National Emergency Management Association (NEMA), and the International Association of Emergency Managers (IAEM), the 2004 edition of the NFPA 1600 *Standard on Disaster/ Emergency Management and Business Continuity Programs*. This coordinated effort was reflected in the expansion of the title of the standard for disaster and emergency management, as well as information on business continuity programs.

SOURCE: NFPA

The NRP provides the structure and mechanisms for the coordination of Federal support to state, local, tribal, and incident managers, and for exercising direct Federal authorities and responsibilities. It assists in the important security mission of preventing terrorist attacks within the United States, reducing the vulnerability to all natural and manmade hazards, and minimizing the damage and assisting in the recovery from any type of incident that occurs.

The NRP is the core plan for managing domestic incidents and details the Federal coordinating structures and processes used during Incidents of National Significance.

The National Incident Management System (NIMS) establishes standardized incident management processes, protocols, and procedures that all responders (Federal, state, local, and tribal) will use to coordinate and conduct response actions. With responders using the same standardized procedures, they will all share a common focus, and will be able to place full emphasis on incident management when a homeland security incident occurs, whether a manmade or natural disaster. In addition, national preparedness and readiness in responding to and recovering from an incident is enhanced because all of the Nation's emergency teams and authorities are using a common language and set of procedures.

Using the NIMS and NRP framework, the shelter plan should implement direction and control for managing resources, analyzing information, and making decisions. The direction and control system described below assumes a facility of sufficient size. Some facilities may require a less sophisticated system, although the principles described here will still apply.

At the Federal headquarters level, incident information-sharing, operational planning, and deployment of Federal resources are coordinated by the Homeland Security Operations Center (HSOC), and its component element, the National Response Co-ordination Center (NRCC).

The national structure for incident management establishes a clear progression of coordination and communication from the local level to the regional level to the national headquarters level. The local incident command structures (namely the Incident Command Post (ICP) and Area Command) are responsible for directing on-scene emergency management and maintaining command and control of on-scene incident operations. Figure 4-2 is a flowchart of initial National-level incident management actions.

A CBRE event can affect a large region and the shelter designer should consider how response and recovery teams can access and work in the vicinity of an incident as shown in Figure 4-3.

An Emergency Management Group (EMG) is the team responsible for the direction and control of a shelter plan. It controls all incident-related activities. The Incident Commander (IC) oversees the technical aspects of the response. The EMG supports the IC by allocating resources and by interfacing with the community, the media, outside response organizations, and regulatory agencies. The EMG is headed by the Emergency Director (ED), who should be the facility manager. The ED is in command and control of all aspects of the emergency. Other EMG members should be senior managers who have the authority to:

- Determine the short- and long-term effects of an emergency

- Order the evacuation or shutdown of the facility

- Interface with outside organizations and the media

- Issue press releases

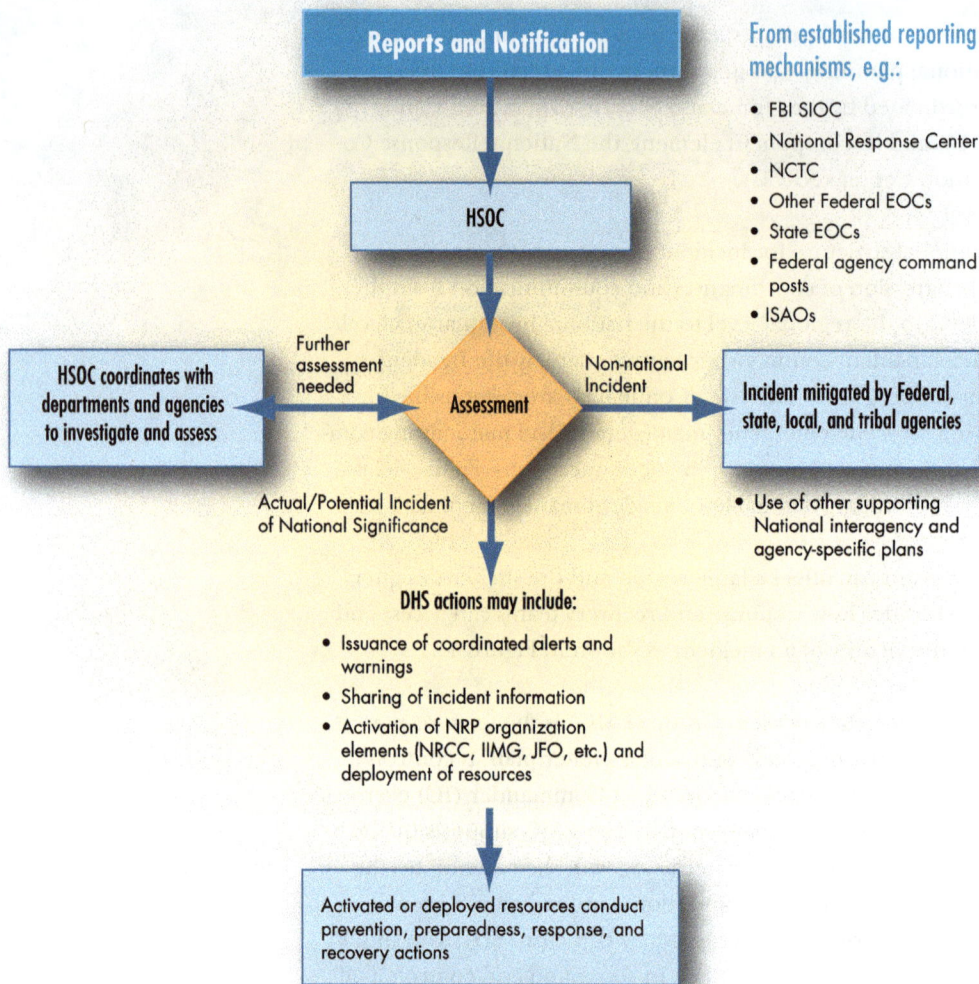

Reports and Notification

From established reporting mechanisms, e.g.:

- FBI SIOC
- National Response Center
- NCTC
- Other Federal EOCs
- State EOCs
- Federal agency command posts
- ISAOs

HSOC

Further assessment needed

HSOC coordinates with departments and agencies to investigate and assess

Assessment

Non-national Incident

Incident mitigated by Federal, state, local, and tribal agencies

- Use of other supporting National interagency and agency-specific plans

Actual/Potential Incident of National Significance

DHS actions may include:

- Issuance of coordinated alerts and warnings
- Sharing of incident information
- Activation of NRP organization elements (NRCC, IIMG, JFO, etc.) and deployment of resources

Activated or deployed resources conduct prevention, preparedness, response, and recovery actions

A basic premise of the NRP is that incidents are generally handled at the lowest jurisdictional level possible. In an Incident of National Significance, the Secretary of Homeland Security, in coordination with other Federal departments and agencies, initiates actions to prevent, prepare for, respond to, and recover from the incident. These actions are taken in conjunction with state, local, tribal, nongovernmental, and private-sector entities.

Figure 4-2   Flowchart of initial National-level incident management actions

SOURCE: DHS NATIONAL RESPONSE PLAN

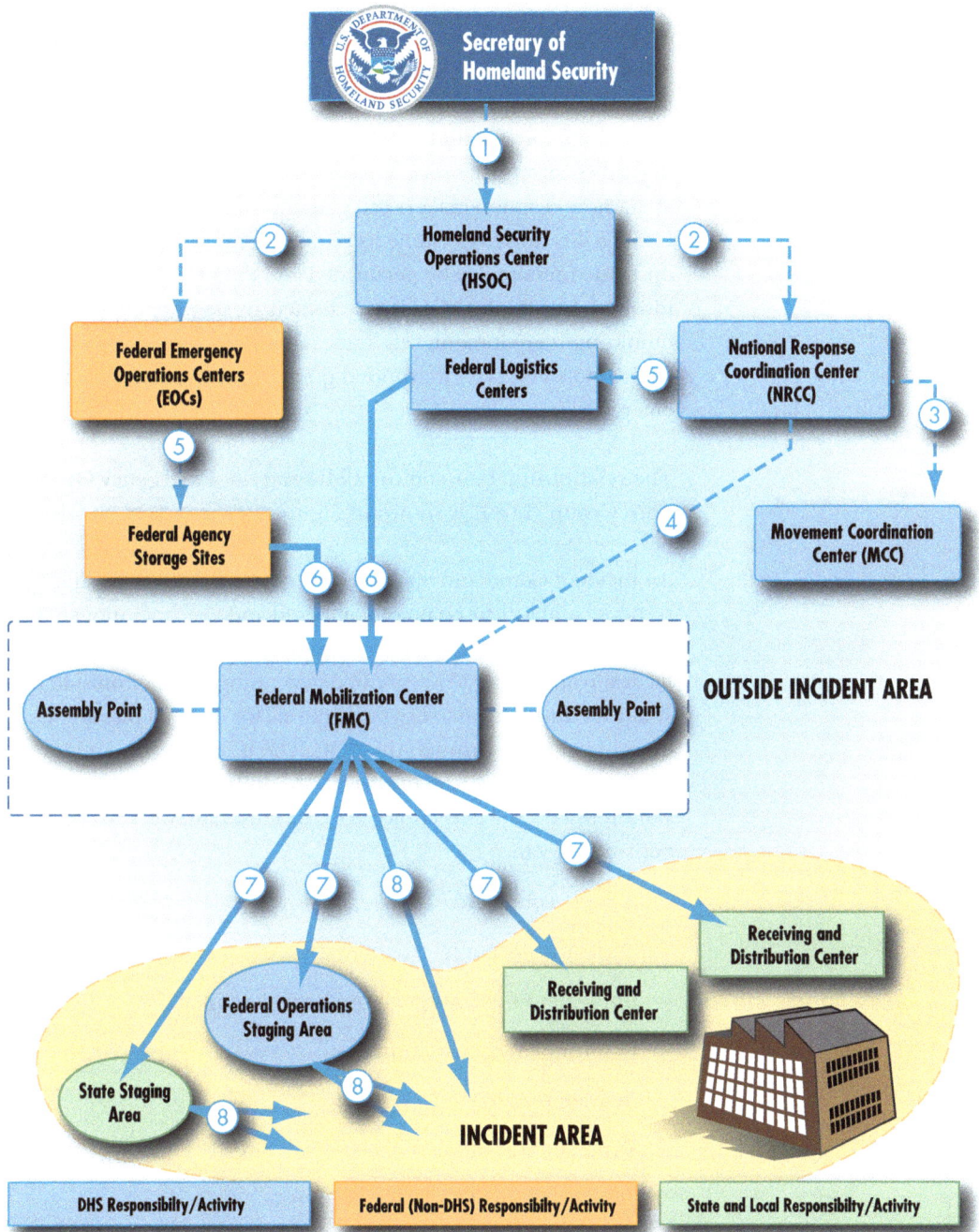

Figure 4-3   NRP-CIS Mass Casualty Incident Response

SOURCE: NRP-CIS

An Emergency Operations Center (EOC) should be established within the shelter that serves as a centralized management center for emergency operations. Here, decisions are made by the EMG based upon information provided by the IC and other personnel. Regardless of size or process, every facility should designate an area where decision-makers can gather during an emergency. Each facility must determine its requirements for an EOC based upon the functions to be performed and the number of people involved. Ideally, the EOC is a dedicated area equipped with communications equipment, reference materials, activity logs, and all the tools necessary to respond quickly and appropriately to an emergency.

The relationship between the EMG and the Emergency Operations Group (EOG) is shown in Figure 4-4.

An Incident Command System (ICS) provides for coordinated response and a clear chain of command and safe operations. The IC is responsible for front-line management of the incident, for tactical planning and execution, determining whether outside assistance is needed, and relaying requests for internal resources or outside assistance through the EOC. The IC can be any employee, but a member of management with the authority to make decisions is usually the best choice. The IC must have the capability and authority to:

○ Assume command

○ Assess the situation

○ Implement the emergency management plan

○ Determine response strategies

○ Activate resources

○ Order an evacuation

○ Oversee all incident response activities

○ Declare that the incident is "over"

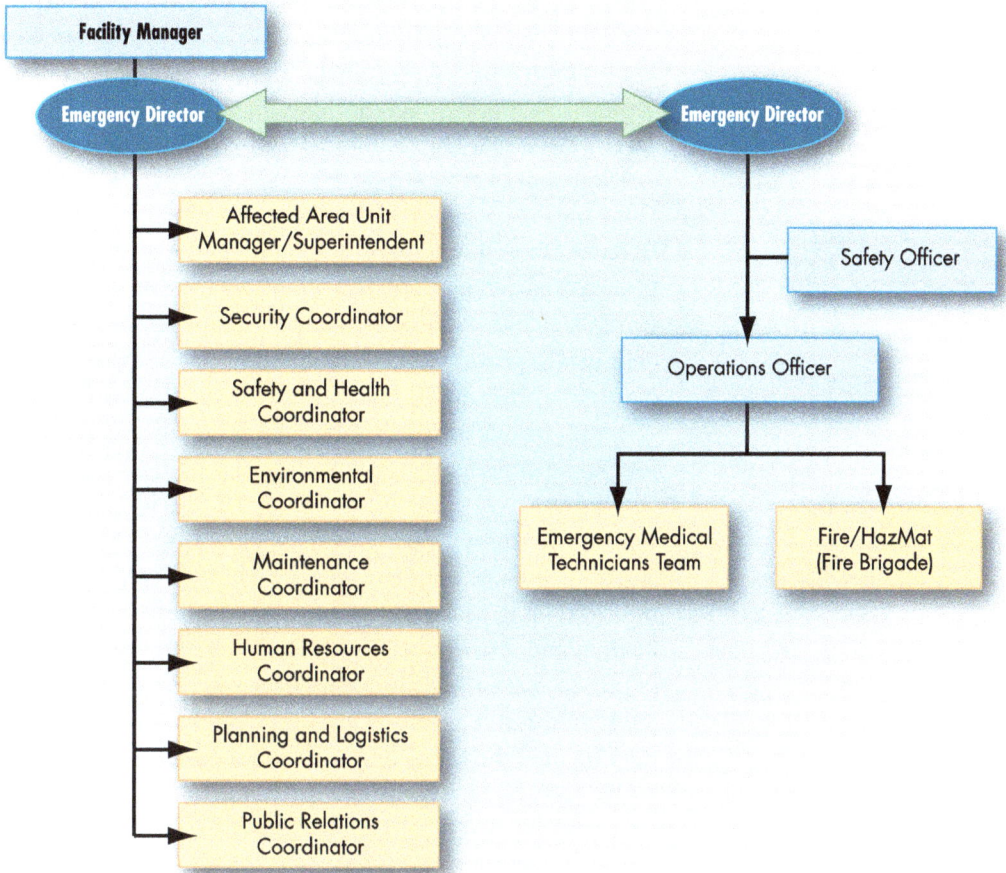

Figure 4-4   Emergency Management Group and Emergency Operations Group

4.3 FEDERAL CBRE RESPONSE TEAMS

The NIMS standardizes resource and asset typing. The following teams are resources that have been typed or are in the process of being typed. These teams are good examples of how to type re-sources, are available for state and Federal response operations, and can provide technical design guidance for shelters:

○ **Weapons of Mass Destruction-Civil Support Team (WMD-CST):** A team that supports civil authorities at a domestic CBRE incident site by identifying CBRE agents/substances, assessing current and projected consequences, advising on response measures, and assisting with appropriate requests for state support. The National Guard Bureau fosters the development of WMD-CSTs.

○ **Disaster Medical Assistance Team (DMAT):** A group of professional and paraprofessional medical personnel (supported by a cadre of logistical and administrative staff) designed to provide emergency medical care during a disaster or other event. The National Disaster Medical System (NDMS), through the U.S. Public Health Service (PHS), fosters the development of DMATs.

○ **Disaster Mortuary Operational Response Team (DMORT):** A team that works under the guidance of local authorities by providing technical assistance and personnel to recover, identify, and process deceased victims. DMORTs are composed of private citizens, each with a particular field of expertise, who are activated in the event of a disaster. The NDMS, through the PHS, and the National Association for Search and Rescue (NASAR) fosters the development of DMORTs.

○ **National Medical Response Team-Weapons of Mass Destruction (NMRT-WMD):** A specialized response force designed to provide medical care following a nuclear, biological, and/or chemical incident. This unit is capable of providing mass casualty decontamination, medical triage, and primary and secondary medical care to stabilize victims for transportation to tertiary care facilities in a hazardous material environment. There are four NMRT-WMDs geographically dispersed throughout the United States. The NDMS, through the PHS, fosters the development of NMRTs.

○ **Urban Search and Rescue (US&R) Task Force:** A highly trained team for search-and-rescue operations in damaged or collapsed structures, hazardous materials evaluations, and stabilization of damaged structures; it also can provide emergency medical care to the injured. US&R is a partnership between local fire departments, law enforcement agencies, Federal and local governmental agencies, and private companies.

○ **Incident Management Team (IMT):** A team of highly trained, experienced individuals who are organized to manage large and/or complex incidents. They provide full logistical support for receiving and distribution centers. Each IMT is hosted and managed by one of the United States Forest Service's Geographic Area Coordination Centers.

4.4 EMERGENCY RESPONSE

Although the NIMS, the NRP, and the National Preparedness Goal provide the designer with factors that may impact shelter management on a regional scale, the incident response occurs at the local level. The IC must evaluate the situation and make a number of time critical decisions. A shelter's location, orientation, and surrounding property adjacent to the site must be evaluated and the locations of the entry access control point, decontamination and disposal areas, and site cordons established, often with little more than the visual inspection of the event area.

4.4.1 General Considerations

The shelter site and surrounding areas should be selected to allow law, fire, and medical vehicles and personnel access for mass decontamination operations in case of an emergency. Runoff from decontamination operations must be controlled or contained to prevent further site contamination. To help the IC, the Emergency Response to Terrorism Job Aid 2.0 should be used. This includes both tactical and strategic issues that range from line personnel to unit officers.

The Job Aid is divided into five primary sections that are tabbed and color coded for rapid access to information:

○ Introduction (Gray)

○ Operational Considerations (Yellow)

○ Incident-Specific Actions (White)

○ Agency-Related Actions (Blue)

○ Glossary (Tan)

As the IC begins the direction of the response and recovery teams in the field, the mobilization of resources to coordinate the Federal, state, local, and tribal efforts will have begun.

4.4.2 Evacuation Considerations

Many of the NIST findings (see Section 1.9) and recommendations for emergency response can be applied to all building types and shelters:

○ Active fire protection systems for many buildings are designed to the same performance specifications, regardless of height, size, and threat profile.

○ Approximately 87 percent of the World Trade Center (WTC) occupants, and over 99 percent of those below the floors of impact, were able to evacuate successfully.

○ At the time of the aircraft impacts, the towers were only about one-third occupied. Had they been at the full capacity of 25,000 workers and visitors per tower, computer egress modeling indicated that a full evacuation would have required about 4 hours. Under those circumstances, over 14,000 occupants might have perished in the building collapses.

○ There were 8,900 ± 750 people in WTC 1 at 8:46 a.m. on September 11, 2001. Of those, 7,470 (or 84 percent) survived, while 1,462 to 1,533 occupants died. At least 107 occupants were killed below the aircraft impact zone. No one who was above the 91st floor in WTC 1 after the aircraft impact survived. This was due to the fact that the stairwells and elevators were destroyed and helicopter rescue was impossible.

○ There were 8,540 ± 920 people in WTC 2 at 8:46 a.m. on September 11, 2001. Of those, 7,940 (or 93 percent) survived, while 630 to 701 occupants were killed. Eleven occupants died below the aircraft impact zone. Approximately 75 percent of the occupants above the 78th floor at 8:46 a.m. had successfully descended below the 78th floor prior to the aircraft impact at 9:03 a.m. The use of elevators and self-initiated evacuation during this period saved roughly 3,000 lives.

○ The delays of about 5 minutes in starting evacuation were largely spent trying to obtain additional information, trying to make sense of the situation, and generally preparing to evacuate.

○ People who started their evacuation on higher floors took longer to start leaving and substantially increased their odds of encountering smoke, damage, or fire. These encounters, along with interruption for any reason, had a significant effect on increasing the amount of time that people spent to traverse their evacuation stairwell.

○ The WTC occupants were inadequately prepared to encounter horizontal transfers during the evacuation process and were occasionally delayed by the confusion as to whether a hallway led to a stairwell as well as confusion about whether the transfer hallway doors would open or be locked.

○ The WTC occupants were often unprepared for the physical challenge of full building evacuation. Numerous occupants required one or more rest periods during stairwell descent.

○ In WTC 1, the average surviving occupant spent approximately 48 seconds per floor in the stairwell, about twice that observed in non-emergency evacuation drills.

The 48 seconds do not include the time prior to entering the stairwell, which was often substantial. Some occupants delayed or interrupted their evacuation, either by choice or instruction.

○ Downward traveling evacuees reported slowing of their travel due to ascending emergency responders, but this counterflow was not a major factor in determining the length of their evacuation time.

○ Approximately 1,000 surviving occupants had a limitation that impacted their ability to evacuate, including recent surgery or injury, obesity, heart condition, asthma, advanced age, and pregnancy. The most frequently reported disabilities were recent injuries and chronic illnesses. The number of occupants requiring use of a wheelchair was very small.

○ Mobility challenged occupants were not universally accounted for by existing evacuation procedures, as some were left by colleagues (later rescued by strangers); some in WTC 1 were temporarily removed from the stairwells in order to allow more able occupants to evacuate the building, and others chose not to identify their mobility challenge to any colleagues.

○ Most mobility challenged individuals were able to evacuate successfully, often with assistance from co-workers or emergency responders, and it is not clear how many were among the 118 from below the impact floors who did not survive. It does not appear that mobility challenged individuals were significantly over-represented amongst the decedents.

○ As many as 40 to 60 mobility challenged occupants and their companions were found on the 12th floor of WTC 1 by emergency responders. About 20 of these were making their way down the stairs shortly before the building collapsed. It is not known how many from this group survived.

❍ The first emergency responders were colleagues and regular building occupants. Acts of individual heroism saved many people whom traditional emergency responders would have been unable to reach in time.

❍ Only one elevator in each building was of use to the responders. To gain access to the injured and trapped occupants, firefighters had to climb the stairs, carrying the equipment with them.

❍ NIST estimated that emergency responder climbing rates varied between approximately 1.4 minutes per floor for personnel not carrying extra equipment to approximately 2.0 minutes per floor for personnel wearing protective clothing and carrying between 50 and 100 pounds of extra equipment.

❍ With a few special exceptions, building codes in the United States do not permit use of fireprotected elevators for routine emergency access by first responders or as a secondary method (after stairwells) for emergency evacuation of building occupants. The elevator use by emergency responders would additionally mitigate counterflow problems in stairwells.

❍ Although the United States conducted research on specially protected elevators in the late 1970s, the United Kingdom along with several other countries that typically utilize British standards have required such "firefighter lifts," located in protected shafts, for a number of years.

❍ Although it is difficult to maintain this pace for more than about the first 20 stories, it would take an emergency responder between 1½ to 2 hours to reach, for example, the 60th floor of a tall building if that pace could be maintained (see Figure 4-5).

High-Rise Buildings and Emergency Response

Example: Fire department response to a 60-story high-rise building, occupants trapped on the 58th and no operating elevators.

Firefighters carrying equipment and wearing PPE ~ 125 minutes

Firefighters carrying no equipment and not wearing PPE ~ 90 minutes

FIRES

— 60th floor
— 58th floor

Firefighters carrying equipment and wearing PPE ~ 70 minutes

Firefighters carrying no equipment and not wearing PPE ~ 50 minutes

— 30th floor

Firefighters begin to climb – 10 minutes

Fire department arrival – 4 minutes

Lobby

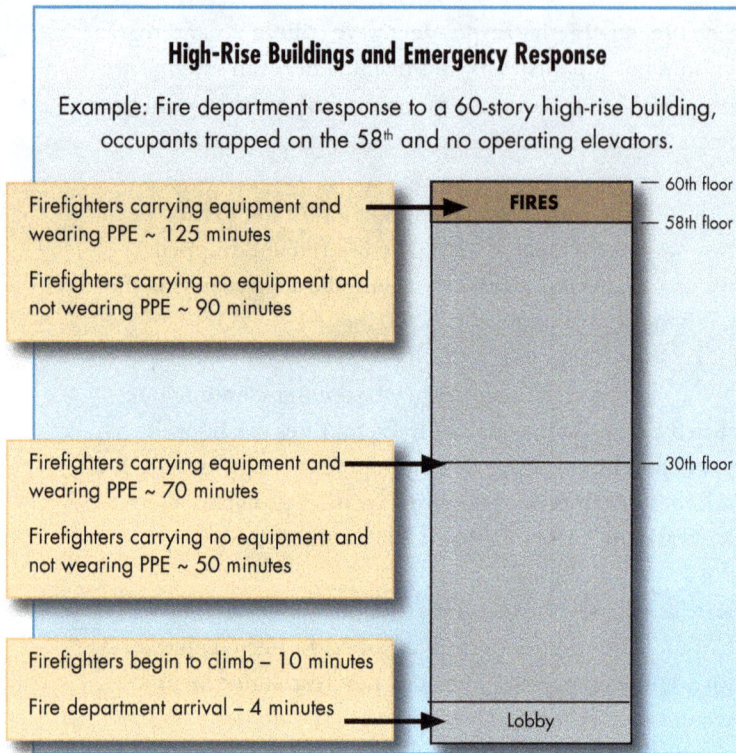

○ Such a delay, combined with the resulting fatigue and physical effects on emergency responders that were reported on September 11, 2001, would make firefighting and rescue efforts difficult even in tall building emergencies not involving a terrorist attack.

4.4.3 Mass Care

The shelter plays a critical role in the mass care and response capability as developed in Appendix 3 of the NRP-CIS:

"Mass Care coordinates Federal assistance in support of Regional, State, and local efforts to meet the mass care needs of victims of a disaster. This Federal assistance will support the delivery of mass care services of shelter, feeding, and emergency first aid to disaster victims; the establishment of systems to provide bulk distribution of emergency relief supplies to disaster victims; and the collection of information to operate a Disaster Welfare Information (DWI) system to report victim status and assist in family reunification. Depending on the nature of the event, a catastrophic disaster will cause a substantial need for mass sheltering and feeding within, near, and beyond the disaster-affected area."

There are a number of assumptions that are used to define the parameters of which the design, utilization, length of occupancy, and shelter capacity should be able to support:

○ As a result of the incident, many local emergency personnel (paid and volunteer) that normally respond to disasters may be dead, injured, involved with family concerns, or otherwise unable to reach their assigned posts.

○ Depending on the nature of the event, a catastrophic disaster will cause a substantial need for mass sheltering and feeding within, near, and beyond the disaster-affected area.

○ State and local resources will immediately be overwhelmed; therefore, Federal assistance will be needed immediately.

○ Extensive self-directed population evacuations may also occur with families and individuals traveling throughout the United States to stay with friends and relatives outside the affected area.

○ Populations likely to require mass care services include the following:

> ○ Primary victims (with damaged or destroyed homes)
>
> ○ Secondary and tertiary victims (denied access to homes)
>
> ○ Transients (visitors and travelers within the affected area)
>
> ○ Emergency workers (seeking feeding support, respite shelter(s), and lodging)

○ In the initial phase (hours and days) of a catastrophic disaster, organized and spontaneous sheltering will occur simultaneously within and at the periphery of the affected area as people leave the area. Additional congregate sheltering may be required for those evacuating to adjacent population centers.

○ The wide dispersal of disaster victims will complicate Federal Government assistance eligibility and delivery processes for extended temporary housing, tracking, and need for registering the diseased, ill, injured, and exposed.

○ More people will initially flee and seek shelter from terrorist attacks involving CBRE agents than for natural catastrophic disaster events. They will also exhibit a heightened concern for the health-related implications related to the disaster agent.

○ Long-term sheltering, interim housing, and the mass relocation of affected populations may be required for incidents with significant residential damage and/or contamination.

○ Substantial numbers of trained mass care specialists and managers will be required for an extended period of time to augment local responders and to sustain mass care sheltering and feeding activities.

○ Timely logistical support to shelters and feeding sites will be essential and required for a sustained period of time. Food supplies from the U.S. Department of Agriculture (USDA) positioned at various locations across the country will need to be accessed and transported to the affected area in a timely manner.

○ Public safety, health, and contamination monitoring expertise will be needed at shelters following CBRE events. Measures to ensure food and water safety will be necessary, and the general public will also need to be reassured concerning food and water safety.

○ Immediately following major CBRE events, decontamination facilities may not be readily available in all locations during the early stages of self-directed population evacuations. People who are unaware that they are contaminated may seek entry to shelters. These facilities may, as a result, become contaminated, adversely affecting resident health and general public trust.

○ Public health and medical care in shelters will be a significant challenge as local EMS resources and medical facilities will likely be overwhelmed quickly. The deployment of public health and medical personnel and equipment to support medical needs in shelters will need to be immediate and sustained by the Department of Health and Human Services (HHS).

○ Shelters will likely experience large numbers of elderly with specific medication requirements and other evacuees on critical home medical care maintenance regimens.

○ Significant numbers of special needs shelters will likely be required as nursing homes and other similar care facilities are rendered inoperable and are unable to execute their evacuation mutual plans and agreements with other local facilities. The American Red Cross will coordinate with HHS in these situations.

4.5 COMMUNITY SHELTER OPERATIONS PLAN

Community shelters should have a Shelter Operations Plan. The plan should describe the different hazards warnings (CBRE, tornadoes, hurricanes, floods, etc.) and Homeland Security Advisory System, and clearly define the actions to be taken for each type of event. A Community Shelter Management Team composed of members committed to performing various duties should be designated. The following is a list of action items for the Community Shelter Operations Plan:

○ The names and all contact information for the coordinators/ managers detailed in Sections 4.5.1 through 4.5.7 should be presented in the beginning of the plan.

○ A hazard event notification, natural or manmade, is issued by the DHS.

○ When an event notification is issued, the Community Shelter Management Team is on alert.

○ When a warning is issued, the Community Shelter Management Team is activated and begins performing the following tasks:

   ○ Sending the warning signal to the community, alerting them to go to the shelter

   ○ Evacuating the community residents from their business or homes and to the shelter

   ○ Taking a head count in the shelter

   ○ Securing the shelter

   ○ Monitoring the event from within the shelter

   ○ After the event is over, when conditions warrant, allowing shelter occupants to leave and return to their homes

   ○ After the event is over, cleaning the shelter and restocking emergency supplies

A member of the Community Shelter Management Team can take on multiple assignments or roles as long as all assigned tasks can be performed effectively by the team member before and during an event.

The following team members would be responsible for overseeing the Community Shelter Operations Plan:

○ Site Coordinator

○ Assistant Site Coordinator

○ Equipment Manager

○ Signage Manager

○ Notification Manager

○ Field Manager

○ Assistant Managers

As previously stated, full contact information (i.e., home and work telephone, cell phone, satellite phone, and pager numbers) should be provided for all team members, as well as their designated backups. The responsibilities of each of these team members are presented in Sections 4.5.1 through 4.5.7. Suggested equipment and supplies for shelters are listed in Section 4.5.8 and Table 4-1.

4.5.1   Site Coordinator

The Site Coordinator's responsibilities include the following:

○ Organizing and coordinating the Community Shelter Operations Plan

○ Ensuring that personnel are in place to facilitate the Community Shelter Operations Plan

○ Ensuring that all aspects of the Community Shelter Operations Plan are implemented

○ Developing community education and training programs

○ Setting up first aid teams

○ Coordinating shelter evacuation practice drills and determining how many should be conducted in order to be ready for a real event

○ Conducting regular community meetings to discuss emergency planning

○ Preparing and distributing newsletters to area residents

○ Distributing phone numbers of key personnel to area residents

○ Ensuring that the Community Shelter Operations Plan is periodically reviewed and updated as necessary

4.5.2   Assistant Site Coordinator

The Assistant Site Coordinator's responsibilities include the following:

○ Performing duties of the Site Coordinator when he/she is off site or unable to carry out his/her responsibilities

○ Performing duties as assigned by the Site Coordinator

4.5.3   Equipment Manager

The Equipment Manager's responsibilities include the following:

○ Understanding and operating all shelter equipment (including communications, lighting, and safety equipment, and closures for shelter openings)

○ Maintaining and updating, as necessary, the Shelter Maintenance Plan (see Section 4.6)

○ Maintaining equipment or ensuring that equipment is maintained year-round, and ensuring that it will work properly during an event

- Informing the Site Coordinator if equipment is defective or needs to be upgraded

- Purchasing supplies, maintaining storage, keeping inventory, and replacing outdated supplies

- Replenishing supplies to pre-established levels following shelter usage

4.5.4   Signage Manager

The Signage Manager's responsibilities include the following:

- Determining what signage and maps are needed to help intended shelter occupants get to the shelter in the fastest and safest manner possible

- Preparing or acquiring placards to be posted along routes to the shelter throughout the community that direct intended occupants to the shelter

- Ensuring that signage complies with ADA requirements (including those for the blind)

- Providing signage in other languages as appropriate for the intended shelter occupants

- Working with the Equipment Manager to ensure that signage is illuminated or luminescent after dark and that all lighting will operate if a power outage occurs

- Periodically checking signage for theft, defacement, or deterioration and repairing or replacing signs as necessary

- Providing signage that clearly identifies all restrictions that apply to those seeking refuge in the shelter (e.g., no pets, limits on personal belongings)

4.5.5 Notification Manager

The Notification Manager's responsibilities include the following:

○ Developing a notification warning system that lets intended shelter occupants know they should proceed immediately to the shelter

○ Implementing the notification system when an event warning is issued

○ Ensuring that non-English-speaking shelter occupants understand the notification (this may require communication in other languages or the use of prerecorded tapes)

○ Ensuring that shelter occupants who are deaf receive notification (this may require sign language, installation of flashing lights, or handwritten notes)

○ Ensuring that shelter occupants with special needs receive notification in an acceptable manner

4.5.6 Field Manager

The Field Manager's responsibilities include the following:

○ Ensuring that shelter occupants enter the shelter in an orderly fashion

○ Pre-identifying shelter occupants with special needs such as those who are disabled or have serious medical problems

○ Arranging assistance for those shelter occupants who need help getting to the shelter (all complications should be anticipated and managed prior to the event)

○ Administering and overseeing first aid by those trained in it

○ Providing information to shelter occupants during an event

○ Determining when it is safe to leave the shelter after an event

4.5.7 Assistant Managers

Additional persons should be designated to serve as backups to the Site Coordinator, Assistant Site Coordinator, Equipment Manager, Signage Manager, Notification Manager, and Field Manager when they are off site or unable to carry out their responsibilities.

4.5.8 Emergency Provisions, Equipment, and Supplies

Shelters designed and constructed to the criteria in this manual are intended to provide short-term safe refuge. These shelters serve a different function from shelters designed for use as long-term recovery shelters after an event; however, shelter managers may elect to provide supplies that increase the comfort level within the short-term shelters. Table 4-1 lists suggested equipment and supplies for community shelters.

Emergency provisions will vary for different hazard events. In general, emergency provisions will include food and water, sanitation management, emergency supplies, and communications equipment. The necessary emergency provisions are as follows:

4.5.8.1 Food and Water. For tornado shelters, because of the short duration of occupancy, stored food is not a primary concern; however, water should be provided. For hurricane shelters, providing and storing food and water are of primary concern. As noted previously, the duration of occupancy in a hurricane shelter could be as long as 36 hours. Food and water would be required, and storage areas for them need to be included in the design of the shelter.

> FEMA and ARC publications concerning food and water storage in shelters may be found at http://www.fema.gov and at http://www.redcross.org, respectively.

4.5.8.2 Sanitation Management. A minimum of two toilets are recommended for both tornado and hurricane shelters. Although the short duration of a tornado might suggest that toilets are not an essential requirement for a tornado shelter, the shelter owner/operator is advised to provide two toilets or at least two self-contained, chemical-type receptacles/toilets (and a room or private area where they may be used) for shelter occupants. Meeting this criterion will provide separate facilities for men and women.

Table 4-1: Shelter Equipment and Supplies

Type	Equipment/Supplies
Communications Equipment	National Oceanic and Atmospheric Administration (NOAA) weather radios or receivers for commercial broadcast if NOAA broadcasts are not available
	Ham radios or emergency radios connected to the police or the fire and rescue systems
	Cellular and/or satellite telephones (may not operate during an event and may require a signal amplifier to be able to transmit within the shelter)
	Battery-powered radio transmitters or signal emitting devices that can signal local emergency personnel
	Portable generators with uninterruptible power supply (UPS) systems and vented exhaust systems
	Portable computers with modem and internet capabilities
	Fax/copier/scanner
	Public address systems
	Standard office supplies (paper, notepads, staplers, tape, whiteboards and markers, etc.)
Emergency Equipment	A minimum of two copies of the Shelter Operations Plan
	Flashlights and batteries, glow sticks
	Fire extinguishers
	Blankets
	Pry-bars (for opening doors that may have been damaged or blocked by debris)
	Stretchers and/or backboards
	Trash receptacles
	Automated External Defibrillator (AED)
	First aid kit
	Trash can liners and ties
	Tool kits
	Heaters
	Megaphones
	Note: many of these items may be stored in wall units or credenzas

Table 4-1: Shelter Equipment and Supplies (continued)

Type	Equipment/Supplies
First Aid Supplies	Adhesive tape and bandages in assorted sizes
	Safety pins in assorted sizes
	Latex gloves
	Scissors and tweezers
	Antiseptic solutions
	Antibiotic ointments
	Moistened towelettes
	Non-prescription drugs (e.g., aspirin and non-aspirin pain relievers, anti-diarrhea medications, antacids, syrup of ipecac, laxatives)
	Smelling salts for fainting spells
	Petroleum jelly
	Eye drops
	Wooden splints
	Thermometers
	Towels
	Fold up cots
	First aid handbooks
Water	Adequate quantities for the duration of the expected event(s)
Sanitary Supplies	Toilet paper
	Moistened towelettes
	Paper towels
	Personal hygiene items
	Disinfectants
	Chlorine bleach
	Plastic bags
	Portable chemical toilet(s), when regular toilets are not contained in the shelter
Infant and Children Supplies	Disposable diapers
	Powders and ointments
	Moistened towelettes
	Pacifiers
	Blankets

Toilets would be needed by the occupants of hurricane shelters because of the long duration of hurricanes. The toilets would need to function without power, water supply, and possibly waste disposal. Whether equipped with standard or chemical toilets, the shelter should have at least one toilet for every 75 occupants, in addition to the two minimum recommended toilets.

4.5.8.3 Emergency Supplies. Shelter space should contain, at a minimum, the following safety equipment:

○ Flashlights with continuously charging batteries (one flashlight per 10 shelter occupants) and glow sticks

○ Fire extinguishers (number required based on occupancy type) appropriate for use in a closed environment with human occupancy, surface mounted on the shelter wall

○ First aid kits rated for the shelter occupancy

○ NOAA weather radio with continuously charging batteries

○ A radio with continuously charging batteries for receiving commercial radio broadcasts

○ A supply of extra batteries to operate radios and flashlights

○ An audible sounding device that continuously charges or operates without a power source (e.g., canned air horn) to signal rescue workers if shelter egress is blocked

4.5.8.4 Communications Equipment. A means of communication other than a landline telephone is recommended for all shelters. Blasts, tornadoes, and hurricanes are likely to cause a disruption in telephone service. At least one means of backup communication should be stored in or brought to the shelter. This could be a ham radio, cellular telephone, satellite telephone, citizens band radio, or emergency radio capable of reaching police, fire, or other emergency service. If cellular telephones are relied upon for communications, the owners/operators of the shelter should install a signal amplifier to send/receive cellular signals from within the shelter. It should be noted that cellular

EMERGENCY MANAGEMENT CONSIDERATIONS

systems may be completely saturated in the hours immediately after an event if regular telephone service has been interrupted.

The shelter should also contain either a battery-powered radio transmitter or a signal-emitting device that can be used to signal the location of the shelter to local emergency personnel should occupants in the shelter become trapped by debris blocking the shelter access door. The shelter owner/operator is also encouraged to inform police, fire, and rescue organizations of the shelter location before an event occurs. These recommendations apply to both aboveground and belowground shelters.

4.5.8.5 Masks and Escape Hoods. Escape hoods, portable air filtration units, and victim recovery units can provide substantial protection and response capability against most agents for a minimal cost and without major changes to the space and structural system.

Escape masks or hoods (personal protective equipment) can be stored at individual desks and in credenzas or wall units in common areas. There are many types of escape masks and hoods that will provide protection against gases and vapors created by fire, chemical and biological agents, and nuclear particles. They can be donned very easily and very fast, generally less than 10 seconds and come in one size fits all.

4.5.8.6 Portable HVAC Units. There are a number of portable filtration units designed for hospital, manufacturing, printing, and other industries that can be used in a safe room with little building modification. The systems typically use HEPA filters to filter the air in a room. Combined HEPA-ultraviolet germicidal irradiation (UVGI) units are now becoming available. These units can provide substantial protection against biological and radiological particulates. There are several units with combined HEPA and activated granular carbon that can provide protection against chemical agents as well. The filtration units can be stored in conference rooms, closets, or in specially designed rooms such as information technology (IT) closets.

4.5.8.7 Emergency Equipment Credenza and Wall Units Storage.
Many Federal government buildings are being outfitted with
either an emergency equipment credenza, or built-in wall
storage units placed in or near the elevator lobby and other
public egress areas. These units can store the first aid kits,
escape hoods and masks, and other emergency preparedness
and response equipment.

4.6 SHELTER MAINTENANCE PLAN

Each shelter should have a maintenance plan that includes the
following:

○ An inventory checklist of the emergency supplies (see Table 4-1)

○ Information concerning the availability of emergency
    generators to be used to provide power for lighting and
    ventilation

○ A schedule of regular maintenance of the shelter to be
    performed by a designated party

Such plans will help to ensure that the shelter equipment and
supplies are fully functional during an event.

4.7 COMMERCIAL BUILDING SHELTER OPERATIONS PLAN

A shelter designed to the criteria of this manual may be used by a
group other than a residential community (e.g., the shelter may
have been provided by a commercial business for its workers or by
a school for its students). Guidance for preparing a Commercial
Building Shelter Operations Plan is presented in this section.

4.7.1   Emergency Assignments

It is important to have personnel assigned to various tasks and
responsibilities for emergency situations before they occur.
An Emergency Committee, consisting of a Site Emergency

Coordinator, a Safety Manager, and an Emergency Security Coordinator (and backups), should be formed, and additional personnel should be assigned to serve on the committee.

The Site Emergency Coordinator's responsibilities include the following:

○ Maintaining a current Shelter Operations Plan

○ Overseeing the activation of the Shelter Operations Plan

○ Providing signage

○ Notifying local authorities

○ Implementing emergency procedures

○ As necessary, providing for emergency housing and feeding needs of personnel isolated at the site because of an emergency situation

○ Maintaining a log of events

The Safety Manager's responsibilities include the following:

○ Ensuring that all personnel are thoroughly familiar with the Shelter Operations Plan and are conducting training programs that include the following, at a minimum:

    ○ The various warning signals used, what they mean, and what responses are required

    ○ What to do in an emergency (e.g., where to report)

    ○ The identification, location, and use of common emergency equipment (e.g., fire extinguishers)

    ○ Shutdown and startup procedures

○ Evacuation and sheltering procedures (e.g., routes, locations of safe areas)

○ Conducting drills and exercises (at a minimum, twice annually) to evaluate the Shelter Operations Plan and to test the capability of the emergency procedures

○ Ensuring that employees with special needs have been consulted about their specific limitations and then determining how best to provide them with assistance during an emergency

○ Conducting an evaluation after a drill, exercise, or actual occurrence of an emergency situation, in order to determine the adequacy and effectiveness of the Shelter Operations Plan and the appropriateness of the response by the site emergency personnel

The Emergency Security Coordinator's responsibilities include the following:

○ Opening the shelter for occupancy

○ Controlling the movement of people and vehicles at the site and maintaining access lanes for emergency vehicles and personnel

○ "Locking down" the shelter

○ Assisting with the care and handling of injured persons

○ Preventing unauthorized entry into hazardous or secured areas

○ Assisting with fire suppression, if necessary

FEMA's United States Fire Administration publication *Emergency Procedures for Employees with Disabilities in Office Occupancies* [http://www.usfa.fema.gov/downloads/pdf/publications/fa-154.pdf] is an excellent source of information on this topic

The Emergency Committee's responsibilities include the following:

○ Informing employees in their assigned areas when to shut down work or equipment and evacuate the area

○ Accounting for all employees in their assigned areas

○ Turning off all equipment

4.7.2 Emergency Call List

A Shelter Operations Plan for a commercial building should include a list of all current emergency contact numbers. A copy of the list should be kept in the designated shelter area. The following is a suggested list of what agencies/numbers should be included:

○ Office emergency management contacts for the building

○ Local fire department—both emergency and non-emergency numbers

○ Local police department—both emergency and non-emergency numbers

○ Local ambulance

○ Local emergency utilities (e.g., gas, electric, water, telephone)

○ Emergency contractors (e.g., electrical, mechanical, plumbing, fire alarm and sprinkler service, window replacement, temporary emergency windows, general building repairs)

○ Any regional office services pertinent to the company or companies occupying the building (e.g., catastrophe preparedness unit, company cars, communications, mail center, maintenance, records management, purchasing/supply, data processing)

○ Local services (e.g., cleaning, grounds maintenance, waste disposal, vending machines, snow removal, post office, postage equipment, copy machine repair)

4.7.3   Event Safety Procedures

The following safety procedures should be followed upon declaration of an event:

○ The person first aware of the event should notify the switchboard operator or receptionist, or management immediately.

○ If the switchboard operator or receptionist is notified, he or she should notify management immediately.

○ Radios or televisions should be tuned to a local news or weather station, and the weather conditions should be monitored closely.

○ If official instruction is given to proceed to shelters or conditions otherwise warrant, management should notify the employees to proceed to and assemble in a designated safe area(s). A suggested announcement would be "The area has been exposed to a CBRE event (type of event). Please proceed immediately to the designated safe area and stay away from all windows."

○ Employees should sit on the floor in the designated safe area(s) and remain there until the Site Emergency Coordinator announces that conditions are safe for returning to work or evacuation.

4.8   GENERAL CONSIDERATIONS

The Shelter Manager and staff should be familiar with how to do the following:

○ Avoid contact with liquids on the ground, victim's clothing, or other surfaces

○ Evaluate signs/symptoms to determine the type of agent involved

○ Separate the victims into groups of symptomatic and asymptomatic, ambulatory and non-ambulatory

○ Prepare occupants for decontamination (patients may undergo gross decontamination with the use of fire hose lines or individual shower and portable decontamination units)

In the case of fire, an immediate evacuation to a predetermined area away from the facility may be necessary. In a hurricane, evacuation could involve the entire community and take place over a period of days. To develop an evacuation policy and procedure:

○ Determine the conditions under which an evacuation would be necessary.

○ Establish a clear chain of command. Identify personnel with the authority to order an evacuation. Designate "evacuation wardens" to assist others in an evacuation and to account for personnel.

○ Establish specific evacuation procedures and a system for accounting for personnel. Consider employees' transportation needs for community wide evacuations.

○ Establish procedures for assisting persons with disabilities and those who do not speak English.

○ Establish post evacuation procedures.

○ Designate personnel to continue or shut down critical operations while an evacuation is underway. They must be capable of recognizing when to abandon the operation and evacuate themselves.

○ Coordinate plans with the local emergency management office.

4.9 TRAINING AND INFORMATION

Employees should be trained in evacuation, shelter, and other safety procedures. Sessions should be conducted at least annually or when:

○ Employees are hired.

○ Evacuation wardens, shelter managers, and others with special assignments are designated.

○ New equipment, materials, or processes are introduced.

○ Procedures are updated or revised.

○ Exercises show that employee performance must be improved.

In addition:

○ Emergency information such as checklists and evacuation maps should be provided.

○ Evacuation maps should be posted in strategic locations.

○ The information needs of customers and others who visit the facility should be considered.

American Concrete Institute. 1999. *Building Code Requirements for Structural Concrete and Commentary.* ACI 318-02, ACI 318-99, and ACI 318R-99. Farmington Hills, MI.

American Red Cross. 2002. *Standards for Hurricane Evacuation Shelter Selection.* ARC 4496. January.

American Society of Civil Engineers, *Minimum Design Loads for Buildings and Other Structures,* ASCE 7-98 Public Ballot Copy, American Society of Civil Engineers. Reston, VA.

American Society for Testing and Materials, *Standard Practice for Specifying an Equivalent 3-Second Duration Design Loading for Blast Resistant Glazing Fabricated with Laminated Glass.* ASTM 2248. ASTM International, 100 Barr Harbor Drive, West Conshohocken, PA.

ANSI/AF&PA NDS-1997. 1997. *National Design Specification for Wood Construction.* August.

Batts, M.E., Cordes, M.R., Russell, L.R., Shaver, J.R. and Simiu, E. 1980. *Hurricane Wind Speeds in the United States.* NBS Building Science Series 124. National Bureau of Standards (NBS), Washington, DC. pp. 41.

Blewett, W.K., Reeves, D.W., Arca, V.J., Fatkin, D.P., and Cannon, B.D. May 1996. *Sheltering in Place: An Evaluation for the Chemical Stockpile Emergency Preparedness Program,* ERDEC-TR-336, U.S. Army Edgewood Research, Development and Engineering Center, Aberdeen Proving Ground, MD.

Blewett, W.K. February 2002. *Fail-Safe Application and Design of Air Conditioners for NBC Collective Protection Systems,* ECBC-TR-223, U.S. Army Edgewood Chemical Biological Center, Aberdeen Proving Ground, MD.

Bureau of Alcohol, Tobacco, Firearms, and Explosives (ATF). 2002. *Incidents, Casualties and Property Damage.*

Blewett, W.K., and Arca, V.J. June 1999. *Experiments in Sheltering in Place: How Filtering Affects Protection Against Sarin and Mustard Vapor,* ECBC-TR-034, U.S. Army Edgewood Chemical Biological Center, Aberdeen Proving Ground, MD.

Carter, R. R. May 1998. *Wind-Generated Missile Impact on Composite Wall Systems.* MS Thesis. Department of Civil Engineering, Texas Tech University, Lubbock, TX.

Clemson University Department of Civil Engineering. January 2000. *Enhanced Protection from Severe Wind Storms.* Clemson University, Clemson, SC.

Coats, D. W., and Murray, R. C. August 1985. *Natural Phenomena Hazards Modeling Project: Extreme Wind/Tornado Hazard Models for Department of Energy Sites.* UCRL-53526. Rev. 1. Lawrence Livermore National Laboratory, University of California, Livermore, CA.

"Design of Collective Protection Shelters to Resist Chemical, Biological, and Radiological Agents." ETL-1110-3-498, February 24, 1999. U.S. Army Corps of Engineers, Washington, DC.

Durst, C.S. 1960. "Wind Speeds Over Short Periods of Time," *Meteorology Magazine,* 89. pp.181-187.

Engelmann, R.J. May 1990. *Effectiveness of Sheltering in Buildings and Vehicles for Plutonium,* DE90-016697, U.S. Department of Energy, Washington, DC.

Federal Emergency Management Agency. 1980. *Interim Guidelines for Building Occupant Protection From Tornadoes and Extreme Winds.* TR-83A. September.

Federal Emergency Management Agency. 1982. *Tornado Protection: Selecting and Designing Safe Areas in Buildings.* TR-83B. October.

FEMA RR-7. 1986. *Civil Defense Shelters A State of the Art Assessment.*

FEMA TR-87. *Standards for Fallout Shelters.*

FEMA TR-29. *Architect and Engineer Activities in Shelter Development.*

Federal Emergency Management Agency. 1988. *Rapid Visual Screening of Building for Potential Seismic Hazards: A Handbook* FEMA 154 Earthquake Hazards Reduction Series 41. July.

Federal Emergency Management Agency. 1988. *Handbook for the Seismic Evaluation of Buildings.* FEMA 310.

Federal Emergency Management Agency. 1997. *NEHRP Recommended Provisions for Seismic Regulations for New Buildings.* FEMA 302A.

Federal Emergency Management Agency. 1999a. *Midwest Tornadoes of May 3, 1999: Observations, Recommendations, and Technical Guidance.* FEMA 342. October.

Federal Emergency Management Agency. 1999b. *National Performance Criteria for Tornado Shelters.* May.

Federal Emergency Management Agency. 2000. *Design and Construction Guidance for Community Shelters.* FEMA 361. July.

Federal Emergency Management Agency. 2003. *Reference Manual to Mitigate Potential Terrorist Attacks Against Buildings.* FEMA 426. December.

Federal Emergency Management Agency. 2004. *Taking Shelter From the Storm: Building a Safe Room Inside Your House.* FEMA 320. March.

Federal Emergency Management Agency. 2004. *Using HAZUS-MH for Risk Assessment.* FEMA 433. August.

Federal Emergency Management Agency. 2005. *A How-To Guide to Mitigate Potential Terrorist Attacks Against Buildings.* FEMA 452. January.

Federal Emergency Management Agency. *Designing for Earthquakes: A Manual for Architects.* FEMA 454. Undated.

Federal Emergency Management Agency and U.S. Fire Administration. *Emergency Procedures for Employees with Disabilities in Office Occupancies.* FEMA 154. Undated.

Fujita, T.T. 1971. *Proposed Characterization of Tornadoes and Hurricanes by Area and Intensity.* SMRP No. 91. University of Chicago, Chicago, IL.

HQ AFCESA/CES, *Structural Evaluation of Existing Buildings for Seismic and Wind Loads.* Engineering Technical Letter (ETL) 97-10.

Kelly, D.L., J.T. Schaefer, R.P. McNulty, C.A. Doswell III, and R.F. Abbey, Jr. 1978. "An Augmented Tornado Climatology." *Monthly Weather Review*, Vol. 106, pp. 1172-1183.

Krayer, W.R. and Marshall, R.D. 1992. *Gust Factors Applied to Hurricane Winds.* Bulletin of the American Meteorology Society, Vol. 73, pp. 613-617.

Masonry Standards Joint Committee. 1999. *Building Code Requirements for Masonry Structures* and *Specification for Masonry Structures.* ACI 530-99/ASCE 5-99/TMS 402-99 and ACI 530.1/ASCE 6-99/TMS 602-99.

Mehta, K.C. 1970. "Windspeed Estimates: Engineering Analyses." *Proceedings of the Symposium on Tornadoes: Assessment of Knowledge and Implications for Man.* 22-24 June 1970, Lubbock, TX. pp. 89-103.

Mehta, K.C., and Carter, R.R. 1999. "Assessment of Tornado Wind Speed From Damage to Jefferson County, Alabama." *Wind Engineering into the 21st Century: Proceedings, 10th International Conference on Wind Engineering*, A. Larsen, G.L. Larose, and F.M. Livesey, Eds. Copenhagen, Denmark. June 21-24. pp. 265-271.

Mehta, K.C., Minor, J.E., and McDonald, J.R. 1976. "Wind Speed Analysis of April 3-4, 1974 Tornadoes." *Journal of the Structural Division, ASCE*, 102(ST9). pp. 1709-1724.

Minor, J.E., McDonald, J.R., and Peterson, R.E. 1982. "Analysis of Near-Ground Windfields." *Proceedings of the Twelfth Conference on Severe Local Storms* (San Antonio, Texas, 11-15 January 1982). American Meteorological Society, Boston, MA.

National Concrete Masonry Association. 1972. *Design of Concrete Masonry Warehouse Walls*. TEK 37. Herndon, VA.

National Fire Protection Association. 1999. *Standard for Healthcare Facilities*. NFPA 99.

National Fire Protection Association. 2003. *Building Construction and Safety Code Handbook*. NFPA 5000.

National Fire Protection Association. 2004. *Disaster/Engineering Management and Business Continuity Programs*. NFPA 1600.

National Fire Protection Association. 2006. *Life Safety Code*. NFPA 101.

NIST Technical Note 1426. U.S. Department of Commerce Technology Administration, National Institute of Standards and Technology, Washington, DC. July.

O'Neil, S., and Pinelli, J.P. 1998. *Recommendations for the Mitigation of Tornado Induced Damages on Masonry Structures*. Report No. 1998-1. Wind & Hurricane Impact Research Laboratory, Florida Institute of Technology. December.

Phan, L.T., and Simiu, E. 1998. *The Fujita Tornado Intensity Scale: A Critique Based on Observations of the Jarrell Tornado of May 27, 1997.*

Pietras, B. K. 1997. "Analysis of Angular Wind Borne Debris Impact Loads." Senior Independent Study Report. Department of Civil Engineering, Clemson University, Clemson, SC. May.

Powell, M.D. 1993. *Wind Measurement and Archival Under the Automated Surface Observing System (ASOS).* Bulletin of American Meteorological Society, Vol. 74, 615-623.

Powell, M.D., Houston, S.H., and Reinhold, T.A. 1994. "Standardizing Wind Measurements for Documentation of Surface Wind Fields in Hurricane Andrew." *Proceedings of the Symposium: Hurricanes of 1992* (Miami, FL, December 1-3, 1993). ASCE, New York. pp. 52-69.

Recommendations to Appendix E, "Planning Guidelines for Protective Actions and Responses for the Chemical Stockpile Emergency Preparedness Program," Section E.4, 17 May 96.

Rogers, G.O., Watson, A.P., Sorensen, J.H., Sharp, R.D., and Carnes, S.A. 1990. *Evaluating Protective Actions for Chemical Agent Emergencies,* ORNL-6615, Oak Ridge National Laboratory, Oak Ridge, TN. April.

Sciaudone, J.C. 1996. *Analysis of Wind Borne Debris Impact Loads.* MS Thesis. Department of Civil Engineering, Clemson University, Clemson, SC. August.

Steel Joist Institute. *Steel Joist Institute 60-Year Manual 1928-1988.* Texas Tech University Wind Engineering Research Center. 1998. *Design of Residential Shelters From Extreme Winds.* Texas Tech University, Lubbock, TX. July.

Twisdale, L.A., and Dunn, W.L. 1981. *Tornado Missile Simulation and Design Methodology.* EPRI NP-2005 (Volumes I and II). Technical Report. Electric Power Research Institute, Palo Alto, CA. August.

Twisdale, L.A. 1985. "Analysis of Random Impact Loading Conditions." *Proceedings of the Second Symposium on The Interaction of Non-Nuclear Munitions with Structures.* Panama City Beach, FL. April 15-18.

U.S. Department of Energy. 1994. *Natural Phenomena Hazards Design and Evaluation Criteria for Department of Energy Facilities.* DOE-STD-1020-94. Washington, DC. April.

General Use Security Documents

DoD Field Manual No. 3-19.30, *Physical Security*, 8 January 2001. Headquarters, Department of the Army, Washington, DC.

UG-2031-SHR: *Protection Against Terrorist Vehicle Bombs*, May 1998. Naval Facilities Engineering Service Center, Security Engineering Division Port Hueneme, CA 93043.

Department of the Air Force, *Force Protection Battlelab Vehicle Bomb Mitigation Guide*, 01 July 1999.

Terrorist Bomb Threat Stand-Off Card. Defense Threat Reduction Agency, Washington, DC. 1995.

Project Development and Design Security Documents

DoD Army TM 5-853-1/AFMAN 88-56, Vol. 1, 5/94, *Security Engineering Project Development*. Department of the Army, U.S. Army Corps of Engineers, Washington, DC 20314-1000.

Army TM 5-853-2/Air Force AFMAN 32-1071, Vol. 2, *Security Engineering Concept Design*, 5/94. Department of the Army, U.S. Army Corps of Engineers, ATM: CEYP-ET, Washington, DC 20314-1OOU.

DoD Army TM 5-853-3/AFMAN 32-1071, Vol. 3, 5/94, S*ecurity Engineering Final Design*. U.S. Army Corps of Engineers, Washington, DC 20314-1OOU.

Army TM 5-853-4, *Security Engineering Electronic Security Systems*, 5/94. U.S. Army Corps of Engineers, Washington, DC 20314-1OOU.

DoD Army TM 5-855-4/AFMAN 32-1071, Vol. 3, 11/86, *Heating, Ventilation, and Air Conditioning of Hardened Installations*. U.S. Army Corps of Engineers, Washington, DC 20314-1OOU.

DoD UFC 4-010-01, *Minimum Antiterrorism Standards*.

DoD UFC 4-023-03, *Design of Buildings to Prevent Progressive Collapse*.

DC 20314-1OOU20314-1OOUB.7 *Army Corps of Engineers Blast Analysis Manual, Part 1 - Level of Protection Assessment Guide*, PDC-TR-91-6 dated 7/91. U.S. Army Corps of Engineers, Omaha, NE.

TDS 2063-SHR, *Blast Shielding Walls*, 9/98. U.S. Army Corps of Engineers, Washington, DC 20314-1OOU.

UG-2030-SHR: *Security Glazing Applications*, 5/98. U.S. Army Corps of Engineers, Washington, DC 20314-1OOU.

Threat, Vulnerability, and Risk Assessment

Risk Assessment Method Property Analysis and Ranking Tool (RAM-PART) being developed by Sandia National Laboratories for GSA. (Currently no copy available)

CNO Antiterrorism/Force Protection Division (N34) Integrated *Vulnerability Assessment (IVA) Guide*.

Vulnerability Assessment Worksheet from U.S. Army Reserve. Headquarters, Department of the Army, Washington, DC.

Port Integrated Vulnerability Assessment (PIVA) For Civilian and Other Non-US Military Ports (Rev-00).

TDS 2062-SHR, *Estimated Damage to Structures from Terrorist Bombs*, 9/98. U.S. Army Corps of Engineers, Washington, DC.

Military Handbook, Design Guidelines for Security Fencing, Gates, Barriers, and Guard Facilities, MIL-HDBK-1013/10, 5/93. Department of Defense, Washington, DC.

Corps of Engineers Guide Specifications for Construction of Progressive Collapse Design Guidance, 4/00. Department of Defense, Washington, DC.

TDS-2079-SHR, *Planning and Design Considerations for Incorporating Blast Mitigation in Mailrooms.* Naval Facilities Engineering Service Center, Port Hueneme, CA.

Corps of Engineers Guide Specifications for Construction of Fencing, 4/99. U.S. Army Corps of Engineers, Washington, DC.

Corps of Engineers Guide Specifications for Construction of Vehicle Barriers, 3/98. U.S. Army Corps of Engineers, Washington, DC.

Corps of Engineers Guide Specifications for Construction of Blast Resistant Doors, 11/97. U.S. Army Corps of Engineers, Washington, DC.

Corps of Engineers Guide Specifications for Construction of Fragment Retention Film for Glass, 7/92. U.S. Army Corps of Engineers, Washington, DC.

Corps of Engineers Guide Specifications for Construction of Security Vault Door, 12/97. U.S. Army Corps of Engineers, Washington, DC.

Corps of Engineers Guide Specifications for Construction of Forced Entry Resistant Components, 4/99. U.S. Army Corps of Engineers, Washington, DC.

Corps of Engineers Guide Specifications for Construction of Bullet-resistant Components, 4/00. U.S. Army Corps of Engineers, Washington, DC.

Corps of Engineers Guide Specifications for Construction of Electromagnetic (Em) Shielding, 4/99. U.S. Army Corps of Engineers, Washington, DC.

Corps of Engineers Guide Specifications for Construction of Self-acting Blast Valves, 7/97. U.S. Army Corps of Engineers, Washington, DC.

Corps of Engineers Technical Letter No. 1110-3-494, *Airblast Protection Retrofit for Unreinforced Concrete Masonry Walls,* 7/99. U.S. Army Corps of Engineers, Washington, DC.

Corps of Engineers Technical Letter No. 1110-3-495, *Estimating Damage To Structures From Terrorist Bombs, Field Operations Guide,* 7/99. U.S. Army Corps of Engineers, Washington, DC.

Corps of Engineers Technical Letter No. 1110-3-498, *Design of Collective Protection Shelters to Resist Chemical, Biological, and Radiological (CBR) Agents,* 2/99. U.S. Army Corps of Engineers, Washington, DC.

Emergency Management and Protective Actions

Preparing Makes Sense. Get Ready Now. http://www.ready.gov

General guidance from DHS on steps to take to prepare for and respond to intentional or accidental releases of chemical, biological and radiological agents and a nuclear blast. Covers schools and daycare, neighborhoods and apartment buildings, and the workplace.

Bioterrorism Preparedness and the Citizen. Centers for Disease Control and Prevention. http://www.pamf.org/bioterror/links.html (chemical and radiological preparedness guidance)

Current guidance on proper actions to take for chemical and radiological events.

Facts About Shelter in Place – Chemical Emergencies. Centers for Disease Control. http://www.bt.cdc.gov/planning/shelteringfacts.pdf

More detailed guidance on protective actions in event of chemical release.

Websites on Sheltering in Place. http://www.scchealth.org/docs/doche/bt/interim.html

Warning, Evacuation and In-Place Protection Handbook, Emergency Management Division, Michigan Division of Emergency Management, 1994. http://floridadisaster.org/bpr/

Warning systems, protective action decision-making, case studies involving chlorine, bromide, and sulfuric acid. Good source for protective actions and shelter considerations in a chemical incident.

Will Duct Tape and Plastic Really Work? Issues Related to Expedient Shelter-In-Place. John Sorensen and Barbara Vogt. August 2001. CSEPP, FEMA.

Defines and discusses expedient sheltering and the effectiveness of select materials, including duct tape.

Shelters by Building Occupancy

Sheltering in the Workplace

Sheltering in Place at Your Office – A general guide for preparing a shelter in place plan in the workplace. National Institute for Chemical Studies. http://www.nicsinfo.org

Provides a sample shelter plan that lists procedures, responsible parties, and needed supplies, equipment and rules.

Fact Sheet on Shelter-in-Place, American Red Cross.

Provides the basics on shelter-in-place at home and the workplace. February 2003

Fire and Explosion Planning Matrix (OSHA, 2004). http://www.osha.gov/dep/fire-expmatrix/index.html

Addresses workplace vulnerability to acts of terrorism and identifies of series of terrorism risk factors (see below) that may elevate the risk of that facility to terrorism acts. These factors may be considered in this project as criteria for higher level of in-place shelter.

In its Worksite Risk Assessment List [http://www.llr.state.sc.us/workplace/sectone.pdf – 507kb PDF], an employer will be asked whether the worksite is characterized by any of the following terrorism risk factors:

- uses, handles, stores or transports hazardous materials;
- provides essential services (e.g., sewer treatment, electricity, fuels, telephone, etc.);
- has a high volume of pedestrian traffic;
- has limited means of egress, such as a high-rise complex or underground operations;
- has a high volume of incoming materials (e.g., mail, imports/exports, raw materials);
- is considered a high profile site, such as a water dam, military installation, or classified site; or
- is part of the transportation system, such as shipyard, bus line, trucking, airline.

Sheltering in Schools

Fairfax County, VA school preparedness and emergency management – good overall document for school preparedness and shelter in place. http://www.fcps.edu/emergencyplan/faq.htm

Comprehensive guidance on protective actions for public schools.

Schools and Terrorism: A Supplement to the National Advisory Committee on Children and Terrorism, Recommendations to the Secretary (August 12, 2003)

Examines the broader issues of integrating school vulnerability and safety issues into community preparedness.

Primer to Design Safe School Projects in Case of Terrorist Attacks. 2003. FEMA. December.

Provides comprehensive guidance to protect students, faculty, staff and their school buildings from terrorist attacks.

Creating a Safe Haven, Dennis Young, http://asumag.com/ar/university_creating_safe_haven/

Guidance on incorporating safe haven principles into school design and construction.

Sheltering in Place – Princeton University

Guidance on protective actions for a campus setting.

Shelters by Hazard – Natural

Hurricanes

Hurricane Shelters, American Red Cross. Provides basic criteria for shelter designation for hurricane shelters. http://www.ih2000.net/jasperem/Hurricane%20-%20Shelters.pdf

Shelter Implementation Workshop. Florida Division of Emergency Management, June 2000.

Proceedings on workshop that addresses problems, issues, and solutions for implementing statewide plan for hurricane shelters.

Standards for Hurricane Evacuation Shelter Selection. American Red Cross (ARC 4496). January 2002

ARC 4496 is the national standard for hurricane evacuation shelter selection criteria. Provides detailed guidance and standards for hurricane shelter selection.

State of Florida Shelter Plan, Florida Division of Emergency Management, 2004. http://floridadisaster.org/bpr/Response/ engineers/documents/2004SESP/Individual%20Elements/2004-SESP-AppxB.pdf

Public shelter design criteria, based on ARC 4496 and Florida design criteria. State requirements for education facilities.

http://floridadisaster.org/bpr/Response/engineers/2004sesp. htm. The website of the Critical Infrastructure and Engineering Unit of the Florida Division of Emergency Management. Contains links to shelter surveys and plans.

Tornadoes and High Winds

Taking Shelter From the Storm: Building a Safe Room Inside Your House. FEMA 320. http://www.fema.gov/pdf/fima/fema320.pdf.

FEMA 320 provides guidance on shelter design and construction of the following types of shelters:

- ⭘ shelter underneath a house
- ⭘ shelter in the basement of a new house
- ⭘ shelter in the interior of a new house
- ⭘ modification of an existing house to add a shelter in one of these areas

FEMA Community Wind Shelters: Background and Research. 2002.

Extreme Event Protection (Hurricanes and Tornadoes). http://www.builtsafe.com/steelclad.pdf

Example of one Texas-based product on the market for extreme wind event protection.

Earthquakes

Federal Emergency Management Agency. 1988. *Handbook for the Seismic Evaluation of Buildings.* FEMA 310. January.

Federal Emergency Management Agency. 1990. *Seismic Considerations for Elementary and Secondary Schools.* FEMA 149.

Federal Emergency Management Agency. 2003. *Existing School Buildings: Incremental Seismic Retrofit Opportunities.* FEMA 318. December.

Shelters by Hazard – Manmade Hazards/Threats

Harden Structures and Systems – Apocalypse House (2003).

Focuses on shelter design for climatic, nuclear, biological, chemical and conventional weapons threats, and the guidelines established by the Federal Emergency Management Agency (FEMA), the U. S. Department of Energy Oak Ridge National Laboratory, as well as more rigorous standards set by the Technical Directives for Shelters by the Swiss Federal Department of Civil Defense.

Building and Shelter Design: Security and Protection Issues

Building Security Through Design: A Primer for Architects, Design Professionals and Their Clients. AIA.

Protecting Occupants of High-Rise Buildings. Rae Archibald (Deputy Fire Commissioner for NYC), http://www.rand.org/publications/randreview/issues/rr.08.02/occupants.html

Recommended actions for building owners of high-rise buildings.

Guidance Publication for Emergency Operations Centers: Project Development and Capabilities Assessment, Florida Division of Emergency

Management (2003) http://floridadisaster.org/bpr/Response/engineers/eoc/eocguide.pdf

Provides guidance on a broad range of vulnerability assessment and vulnerability reduction measures for the FDEM EOC. Many of the recommendations for EOC survivability, sustainability, and interoperability can be applied to multi-hazard shelters.

Security Engineering (Army TM 5-853/Air Force Manual 32-1071)

Design and Analysis of Hardened Structures to Conventional Weapons Effects (TM 5-855-1)

The Homeland Defense Office of the U.S. Army Soldier and Biological Chemical Command (publications, products, and services):

ANSI/ASME N510, *Testing of Nuclear Air Treatment Systems*, 1989.

ASHRAE, *Handbook of Fundamentals*, 1997.

ASHRAE 52.1, *Gravimetric and Dust-Spot Procedures for Testing Air-Cleaning Devices Used in General Ventilation for Removing Particulate Matter*, 1992.

ASHRAE, *Handbook Applications Environmental Control for Survival*, 1982.

ASHRAE Standard 62, *Ventilation for Acceptable Indoor Air Quality*, 1989.

ASHRAE Ventilation Standard 62-1981,1

ASME AG-1, Section FC, *Code on Nuclear Air and Gas Treatment*, 1996.

ASME N509, *Nuclear Power Plant Air-Cleaning Units and Components*, 1989.

ASME NQA-1, *Quality Assurance Requirements for Nuclear Facility Applications*, 1994.

ASTM E779-03, *Standard Test Method for Determining Air Leakage Rate by Fan Pressurization*, 1987.

EA-C-1704, *Carbon-Activated, Impregnated, Copper-Silver-Zinc-Molybdenum-Triethylenediamine (ASZM-TEDA)*, U.S. Army Edgewood Research, Development and Engineering Center (ERDEC), Aberdeen Proving Grounds, MD. January 1992.

ERDEC-TR-336, *Expedient Sheltering In Place: An Evaluation for the Chemical Stockpile Emergency Preparedness Program*, U.S. Army Edgewood Research, Development and Engineering Center (ERDEC), Aberdeen Proving Grounds, MD. June 1996.

FM 3-4, *NBC Protection*, 29 May 1992.

IEEE Std-344, *IEEE Recommended Practice for Seismic Qualification of Class 1E Equipment for Nuclear Power Generating Stations*, 1987.

MIL-PRF-32016(EA), *Performance Specification Cell, Gas Phase, Adsorber*, 26 November 1997.

MIL-STD-282, *Filter Units, Protective Clothing, Gas-Mask Components and Related Products: Performance-Test Method*, 28 May 1956.

MS MIL-F-51079D, *Filter Medium, Fire-Resistant, High-Efficiency*, 17 February 1988.

NFPA 101, *Life Safety Code*, 1997.

TM 5-810-1, *Mechanical Design Heating, Ventilating, and Air Conditioning*, 15 June 1991.

TM 5-855-1, *Design and Analysis of Hardened Structures to Conventional Weapons Effects*, August 1998.

UL 586, *High-Efficiency, Particulate, Air Filter Units*, 1996.

ER 1110-345-100, *Design Policy for Military Construction*.

A

ACI	American Concrete Institute
ADA	Americans with Disabilities Act
AED	Automated External Defibrillator
AF & PA	American Forest and Paper Association
AFCESA	Air Force Civil Engineering Support Agency
AFMAN	Air Force Manual
AIA	American Institute of Architects
ANFO	ammonium nitrate and fuel oil
ANSI	American National Standards Institute
ARC	American Red Cross
ASCE	American Society of Civil Engineers
ASF	anti-shatter film
ASHRAE	American Society of Heating, Refrigeration, and Air-Conditioning Engineers
ASME	American Society of Mechanical Engineers
ASOS	Automated Surface Observing System
ASTM	American Society for Testing and Materials
ATF	Bureau of Alcohol, Tobacco, Firearms, and Explosives

C

CA	California
C&C	components and cladding
CBF	concentric braced frame
CBR	chemical, biological, and radiological
CBRE	chemical, biological, radiological, and explosive
CBRNE	chemical, biological, radiological, nuclear, and explosive
CCA	Contamination Control Area
CCP	Casualty Collection Point
cfm	cubic feet per minute
CIS	Catastrophic Incident Supplement
cm	centimeter
CMU	concrete masonry unit
CO_2	carbon dioxide
CPTED	Crime Prevention Through Environmental Design
CSEPP	Chemical Stockpile Emergency Preparedness Program
CST	Civil Support Team

D

DBT	design basis threat
DC	District of Columbia
DHS	Department of Homeland Security
DMAT	Disaster Medical Assistance Team

DMORT	Disaster Mortuary Operational Response Team
DoD	Department of Defense
DOE	Department of Energy
DOJ	Department of Justice
DWI	Disaster Welfare Information

E

EBF	eccentric braced frame
ECBC	Edgewood Chemical Biological Center
ED	Emergency Director
EDCS	Emergency Decontamination Corridor System
EMG	Emergency Management Group
EMS	Emergency Medical Services
EOC	Emergency Operations Center
EOC	Emergency Operator Coordinator
EOD	Explosive Ordnance Dispersal
EOG	Emergency Operations Group
EPRI	Electric Power Research Institute
ERDEC	Edgewood Research, Development, and Engineering Center (U.S. Army)
ERT-JA	Emergency Response to Terrorism: Job Aid
ESF	Emergency Support Function
ETL	Engineering Technical Letter

F

FBI	Federal Bureau of Investigation
FCO	Federal Coordinating Officer
FDEM	Florida Department of Emergency Management
FEMA	Federal Emergency Management Agency
FF	firefighter
FL	Florida
FMC	Federal Mobilization Center
fps	feet per second
FRC	Federal Resources Coordinator
FRF	fragment retention film
ft	foot, feet
ft²	square foot, feet

G

gal	gallon, gallons
GSA	General Services Administration

H

HazMat	hazardous material
HEGA	high-efficiency gas adsorber
HEPA	high-efficiency particulate air
HHS	Department of Health and Human Services
HPS	Health Physics Society

HQ	Headquarters
hr	hour, hours
HSDL	Homeland Security Digital Library
HSOC	Homeland Security Operations Center
HSPD	Homeland Security Presidential Directive
HSS	Hollow Structural Section
HVAC	heating, ventilation, and air conditioning
IAEA	International Atomic Energy Agency
IAEM	International Association of Emergency Managers
IBC	International Building Code
IC	Incident Commander
ICP	Incident Command Post
ICS	Incident Command System
IEEE	Institute of Electrical and Electronics Engineers
IGU	insulated glazing unit
IIMG	Interagency Incident Management Group
IL	Illinois
IMC	International Mechanical Code
IMT	Incident Management Team
ISAO	Information Sharing and Analysis Organization
ISC	Interagency Security Commitee
IT	information technology
IVA	Integrated Vulnerability Assessment
iwg	inch water gauge

J

JFO   Joint Field Office

K

kb   kilobyte

kg   kilogram

km   kilometer

L

l   liter

lb   pound

lbs   pounds

LDS   Ladder Pipe Decontamination System

LOP   level of protection

LPG   liquefied petroleum gas

M

m   meter

MA   Massachusetts

MCC   Movement Coordination Center

MD   Maryland

MERV   minimum efficiency reporting value

mg   milligram

MI   Michigan

min	minimum
mm	millimeter
mph	miles per hour

N

NASAR	National Association for Search and Rescue
NBS	National Bureau of Standards
NCIS	National Institute for Chemical Studies
NCTC	National Counterterrorism Center
NDMS	National Disaster Medical System
NDS	National Design Specifications
NE	Nebraska
NEHRP	National Earthquake Hazards Reduction Program
NEMA	National Emergency Management Association
NFPA	National Fire Protection Association
NIMS	National Incident Management System
NIST	National Institute of Standards and Technology
NMRT	National Medical Response Team
NMRT-WMD	National Medical Response Team-Weapons of Mass Destruction
NOAA	National Oceanic and Atmospheric Administration
NRCC	National Response Coordination Center
NRP	National Response Plan
NWS	National Weather Service

O

O.C.	on center
OG	outer garments
OH	Ohio
ORNL	Oak Ridge National Laboratory
OSHA	U.S. Occupational Safety and Health Administration

P

PA	Pennsylvania
PAG	Protective Action Guidance
PAO	poly-alpha olefin
PC	personal computer
PDA	personal data assistant
PFO	Principal Federal Official
PHS	U.S. Public Health Service
PIVA	Port Integrated Vulnerability Assessment
PPE	personal protective equipment
PSA	Patient Staging Area
psi	pounds per square inch
PVB	polyvinyl butyral

R

RAMPART	Risk Assessment Method Property Analysis and Ranking Tool

RDD	radiological dispersal device ("dirty bomb")
RRCC	Regional Response Coordination Center

S

SC	South Carolina
SCWB	strong-column weak-beam
SFPE	Society of Fire Protection Engineers
SGP	Sentry Glass ® Plus
SIOC	Strategic Information and Operations Center (FBI HQ)
sq ft	square feet
SRA	Safe Refuge Area
SRWF	shatter-resistant window film
STD	Standard

T

TCL	Target Capabilities List
TFA	Toxic-free Area
TM	Technical Manual
TMS	The Masonry Society
TN	Tennessee
TNT	trinitrotoluene
TR	Technical Report
TX	Texas

U

UBC	Building Code
UCRL	University of California Radiation Laboratory
UFC	United Facilities Criteria
UK	United Kingdom
UL	Underwriters Laboratories
UPS	uninterruptible power supply
URM	unreinforced masonry
U.S.	United States
US&R	Urban Search and Rescue
USDA	U.S. Department of Agriculture
UTL	Universal Task List
UV	ultraviolet
UVGI	ultraviolet germicidal irradiation

V

VA	Virginia

W

WA	Washington
WMD	Weapons of Mass Destruction
WMD-CST	Weapons of Mass Destruction-Civil Support Team
WTC	World Trade Center